T0320403

Space System Architecture Analysis and Wargaming

Space System Architecture Analysis and Wargaming presents a single-source reference for two major topics that are not currently covered in the literature on astronautics.

First, it provides modeling and simulation tools to conduct space system architecting and analysis. These include detailed discussions of various simulation tools, including STK, SEAS, SOAP, AFSIM, EADSIM, and STORM. The book works through the types of analyses that can be performed with each toolkit and focuses on designing and evaluating alternative space systems and system of systems to meet mission requirements. It features chapters written by experts in the field. Second, the text addresses the field of space wargaming. Two tools are addressed, Space Warfare Analysis Tool/Space Attack Warning, and Space and Information Analysis Model. These two tools are addressed by one of the nation's experts in space wargaming.

The book is intended for professionals working in the fields of aerospace engineering, astronautical engineering, space systems engineering, and space wargaming. It will also interest graduate students who are studying spacecraft systems and space architecture.

Space System Architecture Analysis and Wargaming

Edited by Larry B. Rainey

CRC Press
Taylor & Francis Group
Boca Raton London New York

CRC Press is an imprint of the
Taylor & Francis Group, an **informa** business

Designed cover image: Shutterstock

First edition published 2024
by CRC Press
6000 Broken Sound Parkway NW, Suite 300, Boca Raton, FL 33487-2742

and by CRC Press
4 Park Square, Milton Park, Abingdon, Oxon, OX14 4RN

CRC Press is an imprint of Taylor & Francis Group, LLC

Library of Congress Cataloging-in-Publication Data

Names: Rainey, Larry B., editor.
Title: Space system architecture analysis and wargaming / edited by Larry B. Rainey.
Description: First edition. | Boca Raton : CRC Press, Taylor & Francis Group, 2024. |
Includes bibliographical references and index.
Identifiers: LCCN 2023041803 (print) | LCCN 2023041804 (ebook) |
ISBN 9781032343792 (hardback) | ISBN 9781032343808 (paperback) |
ISBN 9781003321811 (ebook)
Subjects: LCSH: Space warfare--Computer simulation. | War games--Data
processing. | Computer architecture. | System analysis.
Classification: LCC UG1530 .S698 2024 (print) |
LCC UG1530 (ebook) | DDC 358/.88--dc23/eng/20240304
LC record available at https://lccn.loc.gov/2023041803
LC ebook record available at https://lccn.loc.gov/2023041804

ISBN: 978-1-032-34379-2 (hbk)
ISBN: 978-1-032-34380-8 (pbk)
ISBN: 978-1-003-32181-1 (ebk)

DOI: 10.1201/9781003321811

Typeset in Times LT Std
by KnowledgeWorks Global Ltd.

Contents

SECTION I   Overview

SECTION II   Space Architecture Tools

SECTION III   Space Wargaming Tools

SECTION IV   Summary

Preface

With the standup of the new Space Warfighting Analysis Center (SWAC) in Colorado Springs, CO, there is the recognition of a new subdiscipline to the field of astronautics as stipulated by the title of this book. The current subdisciplines are Astrodynamics, Spacecraft Propulsion, Spacecraft Design, Control Systems, Space Environment, and Bioastronautics. This text addresses the tools used to perform space system architecture analysis. In addition, this text addresses space wargaming in general and the related subject of space cybersecurity. Moreover, this text addresses two tools to conduct space wargaming specifically.

Dr. Larry B. Rainey, CMSP
Vice President of Engineering
Integrity Systems and Solutions of Colorado, LLC

About the Editor

Dr. Larry B. Rainey is a retired US Air Force space research and development officer. He has over 20 years of experience in the area of space systems engineering. His expertise in the space arena spans working in launch operations, on-orbit operations, space research and development, helping to set up the original Space Warfare Center at Schriever Air Force Base, Colorado, and conducting space systems architecture analysis for The Aerospace Corporation using the following space modeling and simulation tools: Systems Tool Kit(STK), System Effectiveness Analysis Simulation(SEAS), Satellite Orbit Analysis Program(SOAP), Extended Air Defense Simulation(EADSIM), and Synthetic Theater Operations Research Model (STORM). He has also taught spacecraft design at the Air Force Institute of Technology the graduate school of the Air Force, and taught space modeling and simulation at the University of Colorado at Colorado Springs. He has also worked for the Missile Defense Agency's Ballistic Missile Defense System in the Modeling and Simulation Directorate. He has published many articles on systems architecting and applications of cybernetic principles. In addition, he has published six engineering textbooks. One of those is entitled *Space Modeling and Simulation: Roles and Applications Throughout the System Life Cycle.* He has also brought to the attention of the current Space Warfare Analysis Center the relationship between emergent behavior in space system of systems and force design integration as impacted by space threats. He has also consulted with a company supporting Space Systems Command on architecture development.

About the Contributors

John M. Coggi is a Senior Project Leader at the Aerospace Corporation specializing in user-friendly satellite visualization and analysis software applications for desktop computer and web-based environments. In addition to software design and development, Mr. Coggi serves as an orbital analyst for many current and proposed space missions. Mr. Coggi is also the founder of Earthshine Software, LLC. For more information, visit his LinkedIn profile: https://www.linkedin.com/in/john-coggi-984a27b/

Eric Frisco is an expert in the modeling, simulation, and analysis of military space systems. By employing software tools such as the System Effectiveness Analysis Simulation (SEAS), he has worked for over 30 years simulating complex scenarios, predicting system behavior, and evaluating mission effectiveness for the United States Air Force and Space Force. Throughout his 30-year career, Mr. Frisco has primarily supported the modernization of the military space enterprise for the Space and Missile Systems Center, now Space Systems Command, in El Segundo, California. While on active duty in the Air Force he served as the model manager for SEAS and was instrumental in advancing the software to support the developmental planning of future military space systems and led several military utility studies designed to quantify the impact of space services on terrestrial warfighting operations. Today, Mr. Frisco is the National Security Space sector president at ExoAnalytic Solutions, an innovative technology company specializing in space domain awareness, missile defense technology, and modeling and simulation.

David Stodden is a Senior Project Leader in the Engineering Applications Department at the Aerospace Corporation. He develops custom software for orbital simulation and geospatial applications. He is a principal developer of the Satellite Orbit Analysis Program (SOAP). In addition to software development, Mr. Stodden conducts orbital, Radio Frequency (RF), and Terrain analyses using various tools. He has supported numerous studies, including programs such as GPS, CSMAC Working Group 3, and Advanced EHF. David is a U.S. Air Force Veteran and holds an MS Degree in Computer Engineering from California State University, Long Beach.

Paul Szymanski established the Space Strategies Center (SSC) as his private outer space warfare consulting business. His extensive SSC expertise spans 50 years and emphasizes outer space warfare (space control, space superiority, space defense) advanced planning, policy, space military theory, doctrine, strategies, and tactics. This mission area expertise also includes Space Battle Management, Command and Control (BMC2), Space Situational Awareness (SSA) and Space Domain Awareness (SDA), space systems survivability, susceptibility and resilience, space weapons systems effectiveness analyses, autonomous satellite self-defense concepts, and orbital dynamics modeling, simulations, and 3D satellite model construction. Additional expertise includes developing extensive lists (100s) of possible space warfare Courses of Action (COAs) and helping NATO develop their military space policy. In addition, he has a comprehensive experience base, having worked directly with multiple services (Air Force, Army, Navy, Marines), civilian agencies (NASA, DARPA, FEMA), and from the White House National Security Council (NSC) and Congress, to the Pentagon (Secretary of the Air Force) to systems development (Space Systems Command—SSC/ASP/XRJ), technology development (Air Force Research Lab—AFRL/RD/RV/RI/RH and as a member of the Independent Assessment Team general officer programmatic review board) to operational field test (China Lake Naval Test Center). Also, he has published/ lectured in Aviation Week, the United States, the UK, Scotland, Netherlands, France, Germany, Italy, Greece, Estonia, India, Taiwan, Australia, New Zealand, and Japan.

Alexis Wall is an application engineer at AGI, the company that builds Systems Tool Kit (STK). In her role, she focuses on helping users apply STK and other software tools in order to solve their toughest engineering problems using simulation. With an educational and professional background in aerospace engineering, she consults, trains, and provides support to engineers working on space missions in all stages of the life cycle.

Foreword

The threat of war extending to the space domain is a reality that must be taken seriously. As the world becomes more reliant on space-based capabilities, the potential for adversaries to attack those capabilities increases. From communication and navigation to reconnaissance and missile warning, the space domain is integrated into all aspects of warfighting. A disruption of those capabilities could have severe consequences for military operations on the ground, at sea, and in the air.

Moreover, the importance of the space domain in modern warfare cannot be overstated. Space-based capabilities have revolutionized the way wars are fought, and the United States military relies heavily on these capabilities to gain a strategic advantage. Without these capabilities, military operations would be severely hindered, and the ability to project power and defend national interests would be greatly diminished.

The field of astronautics has always been a fascinating and rapidly evolving domain. In recent years, with the establishment of the United States Space Force and the Space Warfighting Analysis Center, the study of astrodynamics in the space domain has become an important area of focus for the warfighting community. As we continue to expand our capabilities in space, it is crucial that we develop new tools and methods for analyzing and understanding this complex environment.

This book, titled *Space Systems Architecture Analysis and Wargaming*, is an essential guide to the tools and techniques that are being used to analyze and model the space domain. It is an important building block in the development within the field of astronautics, with a focus on analysis, modeling, and wargaming.

Moreover, the use of these tools in education and training provides officers with hands-on experience that prepares them for real-world scenarios. By simulating various situations and scenarios, officers can gain practical experience in problem-solving, critical thinking, and decision-making, all of which are essential skills for success in the space domain. Hands-on experience and certifications in the operational domain require training, shadowing, time, and patience. This experience is particularly valuable for future operators as it will provide them with an essential foundation, reducing the time needed to improve operational competencies.

The Air Force Academy's Department of Astronautics is a national leader in developing officers with the skills necessary to understand and operate in the space domain. The use of software modeling and wargaming tools, described in this book, provides students with a unique advantage in their future careers. By providing a strong foundation in these tools and techniques, the Academy is preparing officers to be leaders in the Space Force, capable of tackling complex challenges, and with a deeper understanding of the space domain, allowing them to make more informed decisions and operate more effectively. By using these tools, officers can be enabled to approach their work with a more strategic and nuanced perspective.

The book's five-section organization creates a well-developed background and rationale for understanding and exploring software for space systems architecture analysis and wargaming. The typology of the tools that is used in the space domain provides readers with a framework for understanding the various systems and how they contribute to the understanding of various aspects of operations in the space domain.

Individual chapters in the book focus on a range of space domain modeling and simulation tools critical to both understanding the space domain and how to engage in warfighting operations. Some of these tools include Systems Tool Kit (STK), Satellite Orbit Analysis Program (SOAP), Space Warfare Analysis Tools (SWAT) and Space Attack Warning (SAW), System Effectiveness Analysis Simulation (SEAS), Space and Information Analysis Model (SIAM), Simulation in Python for Space Entities (SPySE), Architecture Framework for Integration Modeling and Simulation (AFSIM), EADSIM, and Synthetic Theater Operations Research Model (STORM). Chapters develop an overview of each package or technique being discussed, explaining its purpose, how it works, and what it can be used for. The chapters also provide real-world examples of how these tools are being used in the space domain, making it easy for readers to see how they can be applied in practice. Together, these tools are currently providing the backbone of warfighting preparation and operations in the space domain.

Overall, the value of modeling and simulation tools for education and development cannot be overstated. Modeling, analysis, and wargaming tools provide officers with a deeper understanding of the space domain, practical experience in problem-solving and decision-making, and a unique advantage in their future careers. As the space domain continues to evolve, the importance of these tools will only continue to grow, and their use will be critical in developing the next generation of leaders in the Space Force.

It is critical that the space domain is protected and defended against potential threats. This requires a comprehensive understanding of the domain and the development of effective tools and strategies for its defense. The modeling and simulation tools described in this book are essential for developing that understanding and preparing for potential threats in the space domain. By providing officers with the knowledge and experience necessary to operate in this complex environment, these tools will play a critical role in ensuring the continued dominance of the military in the space domain.

Space Systems Architecture Analysis and Wargaming is an essential guide for anyone working in the space domain and for those of us educating future generations for those operations. This book provides a comprehensive overview of the tools and techniques that are being used to analyze and model this complex environment, and it will be an invaluable resource for anyone seeking to expand their knowledge and understanding of this exciting and increasingly relevant field.

Luke Sauter, Col, USSF
Professor and Head,
Department of Astronautics
United States Air Force Academy

Section I

Overview

1 Introduction and Overview to Space System Architecture Analysis and Wargaming

Larry B. Rainey

1.1 BACKGROUND/RATIONALE FOR THIS TEXT

During my time in the U.S. Air Force, I worked in the areas of space launch operations and on-orbit space operations. Later, after I retired from the U.S. Air Force, I went to work for The Aerospace Corporation. In this capacity, I taught spacecraft design at the Air Force Institute of Technology, the graduate school for the Air Force, and in a different Aerospace Corporation position I was introduced to the subject of Space System Mission Analysis. I personally ended up using such tools as Systems Tool Kit (STK), Satellite Orbit Analysis Program (SOAP), System Effectiveness Analysis Simulation (SEAS), Extended Air Defense Simulation (EADSIM), and Synthetic Theater Operations Research Model (STORM) for the sake of conducting Space System Mission Analysis. Using such tools as the first three provided me the requisite information needed to be employed in such engagement/mission and campaign tools such as EADSIM and STORM models, respectively, to see the impact of space-derived information at both the engagement/mission and campaign levels of conflict.

The above leads to the conclusion that, like in most professional disciplines, there are subdisciplines to be recognized. The field of astronautics is no exception. The current subdisciplines that constitute the field of astronautics [1] are

- Astrodynamics—the study of orbital motion. Those specializing in this field examine topics such as spacecraft trajectories, ballistics, and celestial mechanics.
- Spacecraft propulsion—how spacecraft change orbits, and how they are launched. Most spacecraft have some variety of rocket engine, and thus most research efforts focus on some variety of rocket propulsion, such as chemical, nuclear, or electric.
- Spacecraft design—a specialized form of systems engineering that centers on combining all the necessary subsystems for a particular launch vehicle or satellite.
- Controls—keeping a satellite or rocket in its desired orbit (as in spacecraft navigation) and orientation (as in attitude control).

DOI: 10.1201/9781003321811-2

- Space environment—although more a subdiscipline of physics than of astronautics, the effects of space weather and other environmental issues constitute an increasingly important field of study for spacecraft designers.
- Bioastronautics—understanding the effect of space on man and how to ensure man can operate in space.

With the publication of this text, there is the rationale and genesis for the instantiation of a new seventh subdiscipline to be added to the above list to be called/labeled Space System Mission Analysis.

1.2   TYPOLOGY OF TOOLS

The Department of Defense (DoD) has a typology or categorization of the various types of tools that are used for modeling, simulation, and analysis of varying scenarios. They are Campaign (force-on-force), Mission (few-on-few), Engagement (one-on-one/few), and Physics (components and subsystem effects). These types of tools are characterized in what is known as the Modeling Pyramid [2]. This is illustrated in Figure 1.1.

In the description of each chapter (i.e., tool in this text), we will see the identification of what type of tool it is.

1.3   WHAT THIS TEXT INVOLVES/WILL ADDRESS

This text will address the following Space System Mission Analysis tools: Systems Tool Kit (STK), Systems Effectiveness Analysis Simulation (SEAS), Satellite Orbit Analysis Program (SOAP), Advanced Framework for Simulation, Integration, and

FIGURE 1.1   Department of Defense (DoD) modeling pyramid.

Modeling (AFSIM), Synthetic Theater Operations Research Model (STORM), and Extended Air Defense Simulation (EADSIM). In addition, two Space Wargaming Tools will also be addressed. They are the Space Warfare Analysis Tool/Space Attack Warning as well as the Space Information Analysis Model.

The following is a description of each tool:

- STK is an indispensable digital mission engineering application for aerospace, defense, telecommunications, and other industries. It features an accurate, physics-based modeling environment to analyze platforms and payloads in a realistic mission context [3].
- SEAS is a constructive modeling and simulation tool that enables mission-level Military Utility Analysis (MUA). Sponsored by the Air Force Space and Missile Systems Center, Directorate of Developmental Planning (SMC/XR), SEAS was created to support developmental planning and Pre-Milestone "A" acquisition decisions for military space systems. SEAS has proven to be a valuable military operations research tool by enabling exploratory analysis of new system concepts, system architectures, and Concepts of Operations (CONOPS) in the context of joint warfighting scenarios [4].
- SOAP is an interactive software system that employs 3D graphics animation to display the relative motion of satellites, airplanes, ships, and ground stations. Users may construct coordinate systems, sensor shapes, and wireframe spacecraft models. A variety of XY plots and data reports may also be generated. SOAP provides analysis, visualization, and simulation capabilities that can answer a variety of questions that commonly arise in space system modeling. This chapter presents an overview of SOAP Version 8, targeted for an early 1995 release on MS-DOS and OSF/Motif based computers. Version 8 offers a rich set of display and analysis features, all integrated into an object-oriented environment with a modern graphical user interface. Propagation of satellites, airplanes/ships, ground stations, the Sun, and the Moon is accomplished in SOAP by using the ASTROLIB astrodynamics routines developed by the Aerospace Corporation. A new C-language implementation of ASTROLIB provides a flexible application programmer interface that is well-suited for the development of interactive software [5].
- AFSIM is an engagement and mission-level simulation environment written in C++ originally developed by Boeing and now managed by the Air Force Research Laboratory (AFRL). AFSIM was developed to address analysis capability shortcomings in existing legacy simulation environments as well as to provide an environment built with more modern programming paradigms in mind. AFSIM can simulate missions from subsurface to space and across multiple levels of model fidelity. The AFSIM environment consists of three pieces of software: the framework itself, which provides the backbone for defining platforms and interactions; an integrated development environment (IDE) for scenario creation and scripting; and a visualization tool called VESPA. AFSIM also provides a flexible and easy-to-use agent modeling architecture, which utilizes behavior trees and hierarchical tasking

called the Reactive Integrated Planning architecture (RIPR). AFSIM is currently ITAR-estricted and AFRL only distributes AFSIM within the DoD community. However, work is underway to modify the base architecture facilitating the maintenance of AFSIM versions across multiple levels of releasability [6].

- EADSIM is a system-level simulation of air, space, and missile warfare developed by the U.S. Army Space and Missile Defense Command Space and Missile Defense Center of Excellence's Capability Development Integration Directorate. EADSIM provides an integrated tool to support joint and combined force operations and analyses. EADSIM is also used to augment exercises at all echelons with realistic air, space, missile, and battle management, command, control, communications, and intelligence (BM/C3I) warfare. EADSIM is used by operational commanders, trainers, combat developers, and analysts to model the performance and predict the effectiveness of ballistic missiles, surface-to-air missiles, aircraft, and cruise missiles in a variety of user-developed scenarios. EADSIM is one of the most widely used simulations in the DoD [7].

- The Synthetic Theater Operations Research Model STORM is the primary campaign analysis tool used by the Office of the Chief of Naval Operations, Assessment Division OPNAV N81, and other DoD organizations to aid in providing analysis to top-level officials on force structures, operational concepts, and military capabilities. This thesis describes how STORM works, analyzes the variability associated with many replications, and evaluates the trade-off between the expected number of replications and the precision and probability of coverage of confidence intervals. The results of this research provide OPNAV 81 with the ability to capitalize on STORM's full potential on a timeline conducive to its high-paced environment. The distribution of outcomes is examined via standard statistical techniques for multiple metrics. All metrics appear to have sufficient variability, which is critical in modeling the combat environment. The tradeoff for confidence intervals between the expected number of replications, precision, and the probability of coverage is very important. If a more precise solution and a higher probability of coverage are required, more replications are generally needed. This relationship is explored and a framework is provided to conduct this analysis on simulation output data [8].

- Space Warfare Analysis Tools/Space Attack Warning (SWAT/SAW): These are software tools that enable the space warfighter to view space object characteristics in a master database; detect pre-attack orbital positioning and evolving attacks in space; task space sensor systems; plan counter-space attacks with standard battle management procedures and orbital dynamics sequencing; visualize the space battlefield, and develop semi-automated Space Wargaming systems.

- Space and Information Analysis Model (SIAM) is a PC-based model under Microsoft Access that tracks and evaluates the flow of information on the battlefield, and its impact on military objectives and command decisions. SIAM analyses can be run from red or blue perspectives and take into account information pathways flowing through both space and terrestrial assets. The SIAM process begins with a strategies-to-task breakdown of military objectives, tasks, force actions, command decisions, and information requirements. Information requirements are linked to space and terrestrial-based sensors, and the flow of critical military information is tracked from these sensors to data processing centers, intelligence centers, command centers, and finally as directives to forces. Every step of this information flow is calculated for the probability of passing to the next step, along with the overall time delay for each step of the process. These information flows can then be denied, delayed, or destroyed, and the effects on the overall probability of the commander receiving this critical information in a timely manner can be assessed. SIAM can be used for both long-term futures planning, and as an operational targeting tool. Both macro and micro views of the battlefield and worldwide sensor data sources can be analyzed. This tool has proven useful in assessing C4I structures, intelligence collection prioritization, possible future space control systems, and information warfare planning.

1.4 MAPPING OF TOOLS ADDRESSED ABOVE TO THE TYPE OF TOOL THAT THEY ARE

Table 1.1 is the mapping of each model addressed above to one of the above mentioned four categories.

1.5 WHAT EACH CHAPTER WILL ADDRESS

Chapter 1 is the introductory chapter for this text. Chapter 11 covers lessons learned and the proposed way ahead.

For Chapters 2, 3, 4, 9, and 10 the following topics will be addressed:

- Title of the chapter (i.e., formal name of the tool to be addressed in the chapter).
- A description of the tool.

TABLE 1.1
Mapping of Tools with the Types of Tools

Tool Name	Type of Tool
System Tool Kit	Mission
SEAS	Mission
SOAP	Mission
Architecture Framework for Simulation, Integration, and Modeling	Mission and Engagement
EADSIM	System Level
STORM	Campaign
SWAT/SAW	Mission and Engagement
SIAM	Mission and Engagement

- An explanation of the various types of analysis that can be performed using the tool.
- An unclassified example of at least one type of analysis that can be performed by the tool.
- Identify what the authors consider to be unique capabilities, or innovative implementations or concepts embraced by their particular tool(s).

For Chapters 5, 6, and 7, there were no available experts who were able to author and disseminate potentially proprietary information on these topics, so I quoted from publicly available fact sheets and any related resources that I could find through online research. For Chapter 7, STORM falls under the International Traffic of Arms Regulations. Therefore, there is no publicly available information (i.e., fact sheet concerning this topic from the DoD or otherwise). Within the above context, STORM and EADSIM are included in the book because these two tools use the information provided by a space asset to assist in conducting an air and ground campaign. AFSIM has its own space capabilities section, which is addressed in Chapter 6.

In summary, for Chapters 2, 3, and 4, there is a requirement to understand the design and operation of a space asset/vehicle to facilitate what information can be provided to a user of the information derived from a space asset. Chapter 5 addresses the space capabilities provided in AFSIM. Chapter 6 provides a fact sheet for EADSIM. Chapter 8 includes all currently available data for STORM. Chapters 9 and 10 address the subject of Space Wargaming and the associated tools for such an investigation.

REFERENCES

1. Subdisciplines to Astronautics. See https://en.wikipedia.org/wiki/Astronautics
2. Modeling Pyramid. See https://www.researchgate.net/figure/Department-of-Defense-Model-Hierarchy-with-Several-US-Air-Force-Exemplar-Models-for-each_fig1_330876622
3. Ansys Space Tool Kit Overview. See https://www.agi.com/products/stk
4. System Effectiveness Analysis Simulation Overview. See https://teamseas.com/about/overview
5. Satellite Orbit Analysis Program Overview. See https://ieeexplore.ieee.org/document/468892
6. Clive, Peter D., Johnson, Jeffrey A., Moss, Michael J., Zeh, James M., Birkmire, Brian M., and Douglas, D. Hodson. "Architecture Framework for Integration, Simulation and Modeling (AFSIM) Overview." International Conference on Scientific Computing, 2015. See worldcomp-proceedings.com/proc/p2015/CSC7058.pdf
7. Extended Air Defense Simulation (EADSIM) Overview. See https://www.tbe.com/missionsystems/eadsim
8. Seymore, Christian. *"Synthetic Theater Operations Research Model (STORM)"*, Naval Post-Graduate School Thesis, 1 September 2014. See https://apps.dtic.mil/sti/citations/ADA621426

Section II

Space Architecture Tools

2 Systems Tool Kit (STK)

Alexis Wall

2.0 INTRODUCTION

Systems Tool Kit is a digital mission engineering tool developed by Ansys Government Initiatives (AGI) for use in aerospace systems modeling and analysis. STK allows users to understand how space, sea, land, and air assets move and interact with each other in their environments. STK is highly configurable for myriad use cases; at its core, it is a tool that enables subsystem, system, and system of systems design and analysis. STK Is accessible through a user interface on Windows machines, where 3D visuals, ground tracks, graphs, and reports are accessible to aid in the user's understanding of systems and system performance. STK can also be run in a no-graphics mode (STK Engine) on Linux or Windows platforms. Its programming interface provides the ability to integrate with external tools and data, build customized tools and workflows, and automate processes. STK supports modeling with dialable levels of fidelity, making it a salient tool for all phases of the life cycle.

2.1 AN OVERVIEW OF STK

Systems Tool Kit (STK) is a COTS (commercial off-the-shelf) software tool that provides a physics-based geometry engine for digital mission engineering and other kinds of analysis. STK models objects in 3D space and over time. It facilitates the modeling of space systems by enabling you to adjust your modeling fidelity, allowing engineers to use high-fidelity models for analysis of the systems they design or operate, and lower fidelity models for lesser known systems. Each subsystem team can work independently on modeling their subsystem, and these models can then be combined into one object in STK. Properties can be updated and changes can be analyzed at the system level at any point in the life cycle. During operations, nominal and off-nominal design reference missions can be modeled to simulate real or potential scenarios.

Design reference missions (DRMs) are scenarios representative of how a model performs in a certain operational context. DRMs can be used to model high-level system behavior or subsystem performance in operational environments of interest. They can be used to evaluate a system's requirements and understand its interactions with related systems. A comprehensive list of DRMs makes up a complete digital mission in the form of an STK mission model.

STK has an extensive API (Application Programming Interface) and can be integrated with many other tools and programming languages. Users can create their own plug-ins for STK to create custom capabilities using built-in functions and available data providers. STK Engine is a graphics-optional version of STK that runs on Windows and Linux systems and can be used to automate scenario generation in STK, build custom applications using STK, or run computations quickly in no-graphics mode.

DOI: 10.1201/9781003321811-4

STK is not only an analytical tool but also a visual tool. It presents reports, plots, and 2D/3D views of the scenario to the user. These visuals are useful for presentations as well as for understanding a system's behavior. It is easier to understand data when a realistic visual representation of the system is available for reference.

2.2   SUBSYSTEM-LEVEL SIMULATION AND ANALYSIS

STK enables users to model the connected architecture of a system of systems. Each subsystem can be modeled at the level of fidelity that is most appropriate to your need and/or the data you have. Simulations can start simply and become more detailed as the design stage progresses. The next subsections will walk through the capabilities of STK for each subsystem on a typical satellite.

2.2.1   ATTITUDE DETERMINATION AND CONTROL

The attitude determination and control subsystem (ADCS) captures the rotational dynamics of a spacecraft. All satellites have sensors that are used to determine the satellite's attitude and actuators to control its attitude. ADCS engineers rely on sensor information to determine a satellite's orientation with respect to the Sun, Earth, and any objects pertinent to its mission. Actuators, such as thrusters and reaction wheels, are used to change the satellite's attitude depending on the task that the satellite needs to perform. Most operational mode changes warrant a change in attitude mode, and the ADCS is responsible for executing the necessary changes.

STK can be used to model dynamic attitude profiles for a satellite. When creating a design reference mission in STK, a user can define the satellite's attitude using common profiles including nadir aligned, Sun aligned, target pointing, precessing spin, and more. There is a multisegment mode that enables users to specify intervals for switching among attitude profiles. Predefined, targeted, and multisegment profiles are extraordinarily useful when investigating how attitude states and state changes affect the system as a whole. For example, the rate of rotation that a satellite uses in safe mode may affect how much sunlight the solar arrays are exposed to over a given interval. In Earth-orbiting missions, solar arrays are critical for supplying power to the spacecraft and keeping it alive. The rotation rate will also affect communications during a pass. If the satellite is spinning too quickly, it may be difficult to lock on to it and acquire signals for long enough to command it to recover. By simulating off-nominal situations such as this in STK, systems engineers and operators can understand how the satellite's attitude affects other subsystems and its ability to conduct the mission.

Though STK provides many built-in attitude profiles for users to assign, the user is also given the option to define their own attitude via an external attitude file. Users can also take advantage of STK's API to script their own attitude simulator and integrate it with STK. This enables ADCS teams to bring their custom or proprietary models into STK for system-level analysis.

For users that need to model more advanced attitude determination and control systems, the Spacecraft Object Library in STK (SOLIS) specialized extension provides numerous options. SOLIS is a commercial plug-in that provides a higher fidelity spacecraft simulation environment within STK. Part of this complete simulation

TABLE 2.1
Attitude Capabilities in SOLIS

Category	Capability
Attitude Disturbance Modeling	Magnetic dipole
	Gravity gradient
	Solar and aerodynamic forces
Attitude Determination Modeling/Guidance	Sensors: sun sensors, horizon sensors, rate sensors, IMUs, magnetometers, star trackers, simulated GPS
	Multiple attitude determination algorithms: perfect attitude determination, fixed-gain filter, Kalman filter
	Orbit determination, ephemeris propagation
Attitude Control Modeling	Actuators: reaction wheels, magnetic torquers, thrusters
	Multiple attitude control algorithms: perfect control, PIID control
Attitude Mode Control	Idle, inertial hold, rate damp
	Tracking and clocking (constraint) modes: Inertial point, Sun track, nadir track, velocity track, mag field track, target tracking, inertial vector tracking, orbit normal track, surface relative velocity track, rotisserie, external guidance, nadir scan, lat/lon scan, Eigen-axis slew, optimized "rendez-slew"

environment is a detailed ADCS simulator. Where STK enables you to model attitude profiles, SOLIS enables you to model how that attitude profile is achieved, maintained, and toggled to support various spacecraft modes. It adds levels of fidelity to ADCS modeling by including the capabilities listed in Table 2.1. One benefit of SOLIS is that it is based on the MAX and ODySSy satellite flight software that is flown on operational satellites. This creates a "simulate like you fly" experience throughout the entire life cycle.

Example

Say you want to model a LEO (Low Earth Orbiter) satellite that scans Earth and images a set of targets, then relays those images down to a ground station. You can begin by defining its attitude system. Sensors are defined for attitude determination and actuators are defined for attitude control. For this example, let's assume the satellite has two types of attitude sensors: Sun sensors and star trackers. The Sun sensors will provide a coarse attitude solution and star trackers will provide a higher fidelity sensed attitude state. The location and orientation of these sensors with respect to the satellite body are defined. Let's assume a simple actuator system that relies solely on momentum biasing. Reaction wheels are added as actuators. A control algorithm is selected and slew modes are defined to use the reaction wheels.

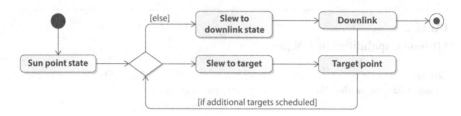

FIGURE 2.1   Attitude states for nominal science collection.

Now that the attitude system is defined, it can be placed into a design reference mission (scenario) and modeled within STK. Imaging targets, as defined in STK, are ingested by SOLIS as a target deck. With user-specified priorities and time constraints, SOLIS's target planner computes a schedule for imaging based on STK access information. This schedule is translated to a sequence with slew/point states that enable the satellite to image all the targets. To remain power-positive, the user may need to schedule a slew to Sun point state before imaging the targets. At the end of the collection period, the satellite must wait for access to the ground station, slew to communicate, then downlink its data. Figure 2.1 represents the attitude states modeled by SOLIS to complete this nominal science collection and downlink procedure.

Figure 2.2 shows how a slew can be visualized in STK. Reaction wheel speeds can be displayed in the 3D graphics window of STK as the SOLIS-defined slew occurs.

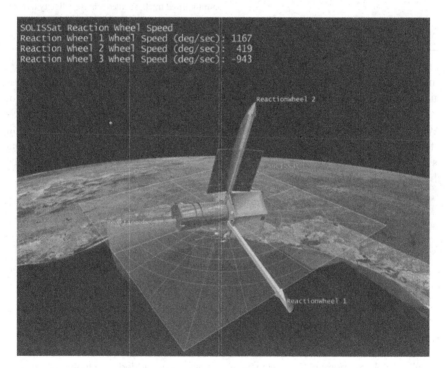

FIGURE 2.2   SOLIS satellite slews with reaction wheels.

2.2.2 FLIGHT DYNAMICS AND GUIDANCE, NAVIGATION, AND CONTROL

A spacecraft's guidance, navigation, and control (GNC) subsystem is responsible for knowing where the satellite is and controlling where it goes. GNC can be broken into two components based on the main tasks the subsystem is responsible for

1. Navigation and orbit determination: GNC teams use information from the satellite's sensors, tracking data on the satellite, and flight dynamics models to understand where the satellite is at a given time.
2. Guidance and control: A satellite's trajectory is carefully defined before launch to meet its mission objective. By the time the satellite separates from its launch vehicle, it is responsible for maneuvering itself to stay on course and adapt to changes prompted by unforeseen circumstances. The GNC team is responsible for computing maneuvers to control the satellite so it can meet its objective.

STK is an incredibly useful tool for navigation and designing orbits. Users can create satellite trajectories or load them using multiple propagators as shown in Figure 2.3. These propagators range in fidelity from a simple two-body propagator to a high-precision orbit propagator with user-defined inputs for gravity models, drag effects, and more. The high-precision orbit propagator (HPOP) is an industry-standard propagator used to simulate satellite trajectories realistically. Additionally, there are other propagators built into STK that handle commonly used data formats for orbit specifications such as two-line element (TLE) files, SPICE kernels, and SP3 files. Though there are many propagators to choose from, there is typically a best choice based on the application. For example, using the high-precision orbit propagator would be excessive when conducting preliminary design. Using the two-body propagator for a mission that traverses cislunar space would be nonsensical due to the gravitational influences of Earth and the Moon.

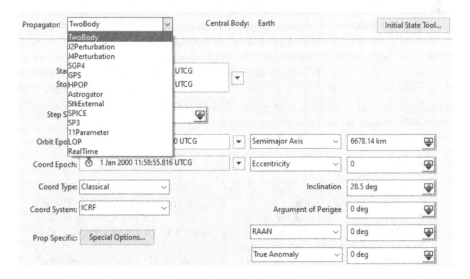

FIGURE 2.3   A satellite's orbit properties and propagator types are displayed.

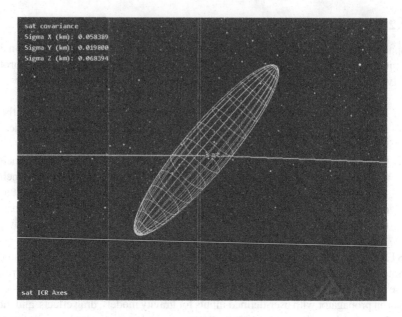

FIGURE 2.4   A satellite's position covariance ellipsoid.

Error in navigation solutions can be incorporated into STK models via covariance information. Covariance is loaded from ephemeris files and visualized in STK as shown in Figure 2.4. To incorporate tracking measurements into the navigation solution and generate covariance information, Ansys's orbit determination tool kit (ODTK) can be used. Section 2.5 contains more information on ODTK.

STK can also be used to design and analyze maneuvers that control a satellite's trajectory via the Astrogator capability. Astrogator enables the user to specify a mission control sequence (MCS)—containing maneuvers, propagate segments (with user-specified propagator types), target sequences, and more—to model all the movement a satellite would undergo over a time period. To perform station-keeping, rendezvous with another satellite, or determine how to get to a lunar orbit in STK, natural dynamics and maneuvers must be leveraged. Within the MCS, users design their trajectories and solve for maneuvers that meet their mission's constraints. The user defines design variables, constraints, and an initial guess, and then Astrogator iterates the design variables to converge on a solution.

Example

You have a low Earth orbiter of interest, and a satellite is being tasked with inspecting it. This inspector satellite must circumnavigate the target satellite and image it. Astrogator is a useful tool for designing rendezvous and proximity operations (RPO) like this. Assuming the inspector satellite starts in the target satellite's orbit, trailing by some in-track offset, Astrogator can be used to determine the necessary maneuvers and periods of propagation for the inspector RPO satellite to approach and circumnavigate the target. Figure 2.5 shows the MCS associated with conducting circumnavigation operations in the target's inertial frame. TheRPO satellite approaches the target satellite then "orbits" it. The target's radial, in-track, and

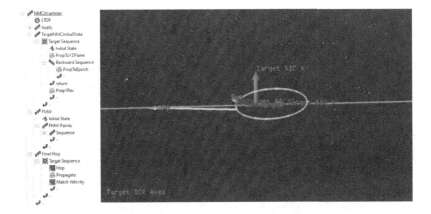

FIGURE 2.5 Inspector satellite (RPO) approaches and circumnavigates target.

cross-track (RIC) vectors are displayed; the RIC frame is useful for analyzing the inspector satellite's location with respect to the target.

2.2.3 Power and Thermal

A satellite's electrical power subsystem (EPS) is responsible for maintaining power positivity and supplying sufficient power to each component of the spacecraft. The thermal subsystem is responsible for maintaining a satellite's temperature so that no components fail in the thermally stressful environment of space. These subsystems are distinct but very dependent upon each other. An average satellite will charge solar arrays to maintain power positivity on orbit. In order to charge solar arrays, the satellite must point them towards the Sun. The Sun's incident angle on various parts of the spacecraft heavily affect its thermal profile. The portion of the satellite exposed to the Sun will heat while the portion in shadow will get very cold. Batteries tend to act like heaters and will keep neighboring components warm when they are active. Likewise, the thermal profile of a satellite will affect noise in measurements across the system.

STK addresses the thermal and power subsystems at a sufficient level for executing early mission design and trade studies. Within STK the space environment, orbit geometry with respect to the Sun/other central body heat sources, and satellite attitude can be modeled. This information is used to compute mean equilibrium temperature of the spacecraft in STK. These data can be pulled from STK and used in other simulation tools to analyze the thermal subsystem. To increase the thermal model's fidelity within STK, SOLIS is used. The satellite bus's heat production, the bus's interaction with each panel on the satellite (conductivity), and thermal properties like absorptivity and emissivity are user defined. By defining these properties and running the simulation, the satellite's thermal trends can be characterized and analyzed. For even higher fidelity, Ansys Thermal Desktop models can be computed using STK's geometry engine.

For power modeling, STK provides a dialable fidelity to understand power production and usage on the satellite. The solar panel tool is used to model the exposure of solar panels mounted on spacecraft, taking self-obscuration of panels and solar incident angle into account.

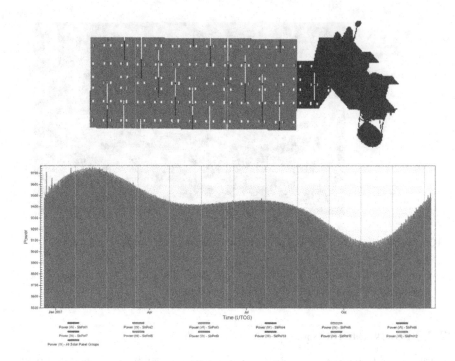

FIGURE 2.6   Power generated by all solar panels on Aqua satellite.

Example

Say you are interested in evaluating a LEO satellite's potential power over a year. Given that a LEO satellite will receive varying solar ray strength over that time, you want to make sure the satellite receives enough sunlight to maintain nominal operations throughout the year. In STK, you can configure your satellite's orbit, nominal attitude profile, and model. By specifying the satellite's solar arrays in the solar panel tool, each panel's power is evaluated for the duration of the simulation. In Figure 2.6, NASA's Aqua satellite is shown with its solar arrays. There are 12 solar panel groups on Aqua that are specified in the solar panel tool. After running the solar panel tool, the combined power generated by all solar arrays is evaluated. It is clear from the plot in Figure 2.6 that the power generated varies throughout the year. In this example, it has the highest power production in February and the lowest in November.

SOLIS can be used to model the power subsystem with higher fidelity. More advanced solar panel properties are accessible to configure; panel disturbances can be modeled, array area fractions (fraction of panel's total area covered by solar cells), and back arrays can be considered when computing solar charging. The loads of various payloads and subsystems can be modeled in SOLIS, and the battery properties can be specified. By setting properties and running SOLIS, various EPS telemetry can be simulated and analyzed using the reporting and graphing features in STK.

2.2.4 COMMUNICATIONS

The communications subsystem, also known as telemetry, tracking, and command (TT&C), acts as the interface between the satellite and the ground. In a simple system, the satellite must be trackable by ground stations, must be able to receive commands from operators on the ground, and must be able to transmit telemetry.

STK is a robust tool for modeling and analyzing communications systems. There are multiple factors that influence communications. At the most basic level, line-of-sight visibility may be required for communications. STK handles geometry and terrain when computing line of sight visibility. This simple definition of access is not sufficient for most communications analyses. An actual communications link must be made between a transmitter and receiver to transfer data. Users can add transmitters and receivers to satellites and ground stations in STK and specify their properties, including antenna patterns. There are several types of transmitter and receiver models in STK that a user can select to model their communications systems as seen in Table 2.2.

TABLE 2.2
Transmitter Models Available in STK

Transmitter Models	Description	Corresponding Receiver Model?
Simple Transmitter	The simplest transmitter type; good for low fidelity analyses. Models an isotropic, omnidirectional antenna and allows you to set a few properties.	✓
Cable Transmitter	For ground stations only: a transmitter that is physically connected to the receiver by some type of "fixed line medium," such as wire, coaxial cable, twisted-pair cable, CAT-5, or fiber optic silica.	✓
Medium Transmitter	Provides more flexibility by letting you specify gain and power separately instead of entering their product (EIRP) directly, as in the Simple model.	✓
Complex Transmitter	Enables you to model many analytical and realistic antenna models and gives you freedom to set all properties.	✓
Multibeam Transmitter	The multibeam transmitter model lets you set up multiple antenna beams, each with its own specs and its own polarization and orientation properties.	✓
Plugin Transmitter	Enables you to use external scripts, such as MATLAB, VBScript, and Perl scripts, to define transmitters for use in your communications link analyses.	✓
Laser Transmitter	The laser transmitter model models the transmit terminal of a laser communications link. In combination with the laser receiver model, it facilitates communications laser link budget analysis.	✓
GPS Satellite Transmitter	Specific values applied to the multibeam transmitter model to emulate a GPS transmitter.	
Simple Retransmitter	Same as simple transmitter but built for retransmitting. Allows you to set transfer functions.	
Medium Retransmitter	Same as medium transmitter but built for retransmitting. Allows you to set transfer functions.	
Complex Retransmitter	Same as complex transmitter but built for retransmitting. Allows you to set transfer functions.	

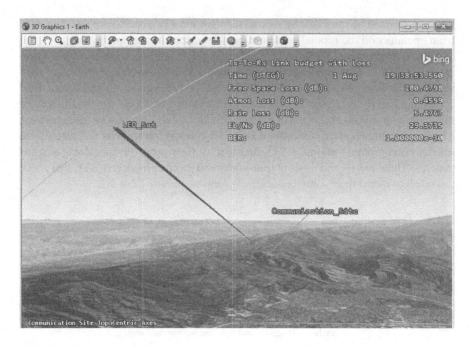

FIGURE 2.7   Link analysis including losses.

To dial up the fidelity even further, users can simulate the radio frequency (RF) environment. Incorporating rain and cloud models, atmospheric absorption, ionospheric propagation fading, and other loss models into the scenario, communications links are further constrained. Example losses associated with the RF environment are displayed in Figure 2.7.

By modeling the environment and a satellite's antennas in STK, you can analyze how the communications subsystem performs. This is not only useful for designing systems, but is also useful for understanding off-nominal events during operations. Custom communications components can be "plugged in" to STK to simulate real systems as accurately as possible, enabling you to evaluate your system before and after launch using STK.

Additionally, communications modeling and analysis can be scaled up in STK. Entire constellations of satellites can communicate with networks of ground stations in STK. Multi-hop satellite links can be modeled and analyzed for relay efficiency.

Example

Aircraft rely on global navigation satellite systems (GNSS) such as GPS to navigate. Say you are interested in understanding how enemy jammers could affect an aircraft's GPS reception while flying a route through treacherous territory. In STK, interference can be modeled using a Comm System object. The GPS constellation's transmitters send signals that are collected by the aircraft's receiver. The

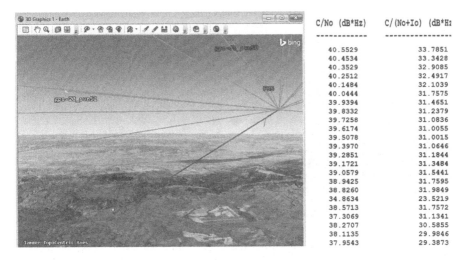

C/No (dB*Hz)	C/(No+Io) (dB*H₂
40.5529	33.7851
40.4534	33.3428
40.3529	32.9085
40.2512	32.4917
40.1484	32.1039
40.0444	31.7575
39.9394	31.4651
39.8332	31.2379
39.7258	31.0836
39.6174	31.0055
39.5078	31.0015
39.3970	31.0646
39.2851	31.1844
39.1721	31.3484
39.0579	31.5441
38.9425	31.7595
38.8260	31.9849
34.8634	23.5219
38.5713	31.7572
37.3069	31.1341
38.2707	30.5855
38.1135	29.9846
37.9543	29.3873

FIGURE 2.8   A jammer interferes with an aircraft's GPS receiver.

aircraft will also receive an interfering signal from a jammer (transmitter) on the ground. Systems like these are illegal in the U.S. but could be used by adversaries during conflict. The jammer, if successful, could invalidate the aircraft's GPS positioning information. In STK, the effects of a GPS jammer on a communications component can be analyzed as shown in Figure 2.8.

A lower carrier-to-noise ratio is observed in this system when interference from a jammer is included.

2.2.5 PAYLOADS

Payloads are the reason satellites are launched. They are responsible for collecting the data or performing the necessary tasks to fulfill a satellite's mission. Satellites may be tasked with capturing images or collecting scientific data to relay to Earth.

STK models a few types of payloads. STK models multiple radar types and modes. Radar data is used to detect and track objects. For advanced radar modeling and analysis, the multifunction radar (MFR) type can be used instead of monostatic or bistatic radar types. The MFR models multiple radar beams working together at the same location, each with its own power specs and constraints. STK also enables users to select from two radar modes: synthetic aperture radar (SAR) or search/track. SARs are often used on satellites because of their ability to achieve high resolution in the cross-range dimension by taking advantage of the motion of the satellite. The search/track radar mode detects and tracks point targets. In STK, a user can set up design reference missions (DRMs) containing target configurations and environment specifications then analyze how their radar system performs. These DRMs can be evaluated to determine if mission objectives are met.

Imaging payloads are often modeled by sensors in STK. For high-level analyses where camera quality does not need to be considered, a user can specify the

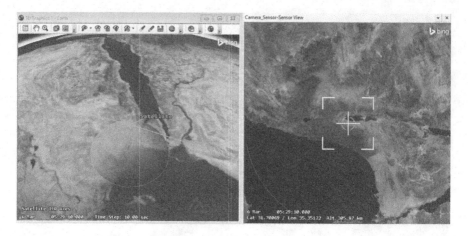

FIGURE 2.9 A satellite has a nadir-pointed sensor with a 90-degree cone. The sensor's view is shown to the right.

sensor shape and attach the sensor to the satellite. A LEO satellite is shown with a 90-degree conic sensor pointed nadir in the left-most image of Figure 2.9. The image on the right shows the sensor scene.

To increase the fidelity of sensor scenes, STK's electro-optical infrared (EOIR) capability is used. This capability models the detection, tracking, and imaging performance of EO and IR sensors. Sensor properties are defined in terms of bands: multiple bands can be considered and each one enables the user to set spatial, spectral, optical, and radiometric properties. After configuring a sensor, objects of interest in the scenario are given EOIR properties. In conjunction with sensor properties, these EOIR 'shape' properties dictate how well the EO/IR sensor will detect objects of interest. STK's EOIR capability also enables the user to define the EOIR configuration/environment. Users can load in atmospheric models, cloud models, or custom texture maps to define the environment in which the EO/IR sensor operates.

Example

You are interested in imaging a satellite of interest from a nearby orbit. If you can successfully image the satellite of interest using your satellite's optical payload, you can gain valuable insight into the object of interest's operations. You configure your satellite to track the object of interest. Using STK's EOIR capability, you set the spatial, optical, and radiometric properties of an optical sensor. The satellite of interest's EOIR shape is defined and the EOIR configuration is set to compute background noise from the Earth and stars. All this information is combined to generate a synthetic sensor scene that updates with time as shown in the bottom left quadrant of Figure 2.10.

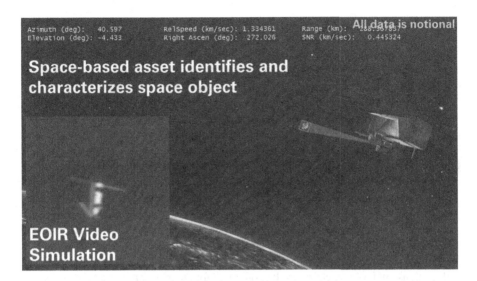

Azimuth (deg): 40.597   RelSpeed (km/sec): 1.334361   Range (km): All data is notional
Elevation (deg): -4.433   Right Ascen (deg): 272.026   SNR (km/sec): 0.445324

Space-based asset identifies and characterizes space object

EOIR Video Simulation

FIGURE 2.10   A synthetic sensor scene is shown for a space-to-space optical payload.

2.3 SYSTEM-LEVEL MODELING AND ANALYSIS IN STK

After modeling individual subsystems for a satellite, a system-level view must be taken. Each subsystem is given requirements and constraints to operate within. Despite engineers' best efforts to build a perfect subsystem, issues will always be discovered during and after integration, and often during operations. The key to efficiently discovering and correcting issues at the system level is to have an accurate model of the system and subsystem interdependencies throughout the satellite's life cycle.

STK can be used as a validation tool throughout a satellite's life cycle. In early design and concept development, STK is used to ensure the mission's concept is achievable. As the design is refined, the fidelity of subsystem modeling is increased and seamlessly integrated into one satellite system. During the test phase, design reference missions place the system into simulated operational environments. When issues arise in the simulated reference missions, they can be addressed on the physical system by altering flight software or ground system architecture. As a satellite flies, nominal and off-nominal operations are simulated in STK to determine if and how the system's operations need to be updated to meet mission requirements.

STK is a system-focused tool with many system-level capabilities. A few are discussed in this section.

2.3.1 COVERAGE EVALUATION AND STK ANALYZER

A satellite system is evaluated by its ability to meet its requirements and fulfill its objectives. Every mission will be required to observe and/or communicate with ground/space-based objects. To evaluate how well a system performs its mission, it must be observed in every configuration and environment of interest.

As engineers and analysts, various mission modes and configurations are identified in early concept development. The satellite's orbit regime will dictate its environment. These mission modes and environments are combined to create various design reference missions that, when combined, represent the entire system. A single DRM could represent the satellite operating in safe mode for a day while operators work to recover it to nominal operations. Another DRM could represent a nominal cycle of data collection, solar panel charging, then communicating data to various ground stations. For each reference mission, the system needs to be evaluated based on different quality metrics. During nominal operations, one quality metric may be how much time a satellite spends imaging its intended target or area of interest. During safe mode, operators are interested in a quality metric pertaining to how much time the satellite's panels spend pointing toward the Sun and charging the satellite's batteries. These quality metrics can be expressed using STK's coverage capability. A satellite is assigned to cover certain assets or areas of interest and quality metrics are defined. By computing coverage, the quality metrics are evaluated. Users can understand a lot about their system by evaluating coverage. Area covered, revisit time, communications link quality, custom calculations, and more metrics can be computed for an analyst to view.

When a system's configuration is not fully known, it can be useful to understand the trade space surrounding system performance and unknown variables. STK Analyzer can be used to explore the design space of a system by producing parametric studies, carpet plots, design of experiments tests, probabilistic analyses, and optimization algorithms. STK Analyzer is a plug-in that makes ModelCenter's trade study capabilities available within STK.

Example

You are doing initial orbit design for a constellation of LEO CubeSats that will collect imagery of Earth. The system's area of interest is the Northern hemisphere and line-of-sight visibility is assumed to be sufficient for imaging at this level of analysis. In STK, a LEO seed satellite is defined and a coverage definition of the Northern hemisphere is defined to evaluate simple coverage. Simple coverage is the percentage of the area of interest covered during the analysis interval. For this example, the analysis interval will cover one orbit period. You are interested in analyzing how the size of the constellation will affect coverage. You can set up a carpet plot in STK Analyzer to vary the number of planes and satellites per plane in a walker constellation and evaluate simple coverage as an output. Walker constellations with two to six planes and two to six satellites per plane are considered. Figure 2.11 shows the relationship between the number of planes/satellites per plane and coverage.

Based on this trade study, it is evident that the number of planes is more influential than satellites per plane when evaluating coverage. The largest effect is observed between two-plane and three-plane constellations. There is a point of diminishing returns around the four to five satellites per plane and four-planes design that can be considered when defining the constellation. This study can be used to drive design decisions and can be integrated with cost models to determine which design is the most practical to execute given cost constraints.

STK Analyzer can be used to vary any system or subsystem inputs and quantify system-level effects.

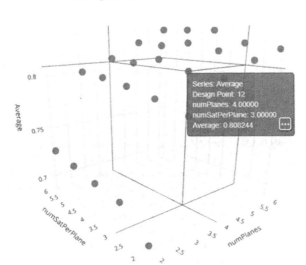

FIGURE 2.11   3D scatter plot showing constellation design's effect on coverage.

2.3.2 SPACE ENVIRONMENT AND EFFECTS

Once launched, a satellite must operate in the harsh conditions of space. Thermal radiation, ionizing particles, and space debris can end a mission if the system is unable to manage these effects.

STK's Space Environment and Effects (SEET) capability models the space environment and its effects on a spacecraft using industry-standard models. These features are listed in Table 2.3.

2.3.3 OPERATOR'S TOOLBOX

When moving out of the realm of design and verification and into the realm of operations, a satellite requires different types of analyses to maintain operability. Operators are interested in analyzing existing systems and understanding how the mission can best be performed daily.

To better serve operators using STK, Operator's Toolbox (OT) was built. It packages existing functionality from STK into simpler, easy-to-use workflows that operators often use. OT is a utility within STK that quickly enables users to

- Load in templates containing pre-defined system configurations. A template may contain one or many satellites, ground assets, targets regions, payloads, and communication components.
- Create objects through a simplified workflow containing only those properties that are most necessary.
- Evaluate passive safety of a satellite during close-proximity operations.

TABLE 2.3
SEET Features

SEET Feature	Description
Radiation Environment	SEET computes the expected dose rate and total dose due to energetic particle fluxes for a range of shielding thicknesses and materials, and can also compute the energetic proton and/or electron fluxes for a wide range of particle energies. This is important for analyzing how components of the vehicle will degrade over time.
South Atlantic Anomaly (SAA)	The SAA is a region of space with an enhanced concentration of ionizing radiation due to the configuration of Earth's magnetic field. Such radiation can damage spacecraft electronics and cause Single Event Upsets (SEU), which can impair the functioning of electronic components. A key feature of SEET is its ability to compute a spacecraft's entrance and exit times through the SAA. The SAA transit component also computes the energy flux and/or flux contour of SEU relative probability for altitudes between 400 and 1500 km.
Galactic Cosmic Ray (GCR)	Multiple GCR models are available to compute differential and integral fluxes and fluences. GCRs can lead to single-event effects and satellite anomalies affecting electronic components and software.
Solar Energetic Particle (SEP)	SEP probabilistic fluences are computed over mission lifetime. SEPs can cause single-event effects as well as increase ionizing dose, leading to long-term failure of electronics and solar arrays.
Particle Impacts	This feature computes the total mass distribution of meteor and orbital debris particles that impact a spacecraft along its orbit during a specified period. It can also compute the mass distribution of these particles above a satellite surface damage threshold you specify. SEET enables you to define or select from lists of surface materials and properties that may be damaged by high-velocity impact with meteors and orbital debris.
Vehicle Temperature	For vehicle subsystem design and operations, thermal environment energy combined with any internal heat dissipation requirements must be considered. Using thermal balancing equations, SEET determines the mean temperature of a space vehicle due to direct solar flux, reflected and infrared Earth radiation, and the dissipation of internally generated heat energy.
Magnetic Field	SEET uses a highly customizable set of conditions to compute the local magnetic field at the current location.

- Compute solar phase (beta) angle between an observer and any number of targets. Solar phase angle dictates how well-lit the target appears relative to the observer and may be constrained using this panel. The definition of solar phase angle is shown in Figure 2.12.
- Calculate when any number of objects cross the orbital plane of a specified satellite. This is useful for collision avoidance and situational awareness.
- Display a sensor boresight's view as shown in Figure 2.9.

- Simulate debris generation in the event of a collision.
- Simulate tracking observations for any space-based asset, process them, and return the results into STK through OT's orbit determination (OD) simulator. This is accomplished by creating a closed-loop simulation between STK and the orbit determination tool kit (ODTK) as shown in Figure 2.13. ODTK will be further described in Section 2.5 It is worth noting that OD simulator can process through maneuvers.

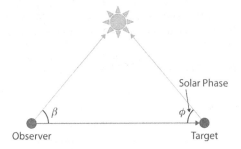

FIGURE 2.12   Solar phase and beta angle.

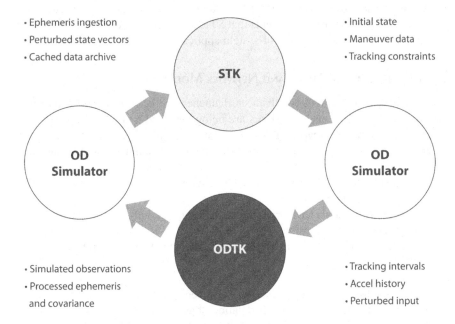

FIGURE 2.13   OD simulator's closed-loop diagram.

2.3.4 Test and Evaluation Tool Kit

It is critical to test a system to understand how it will perform operationally. Prior to launching, a satellite is subjected to various types of tests including environmental tests, operational readiness tests, and more. During operations, tests will be run using simulators and engineering models on the ground prior to executing new sequences or tasks.

Test data require careful scrutinization to understand how the satellite will respond to its stimuli. STK's test and evaluation tool kit (TETK) enables users to load real or simulated data into STK so its analysis and visualization tools can be utilized. This data can come from physical systems or software simulators (such as custom simulators, STK, and ODTK). TETK will interpolate between known data points to provide a continuous time model of the test's data. TETK can be used to understand why communications drop-out occurred during a test, why thrusters fired in the wrong direction, why solar arrays are not charging the expected amount, and more.

2.4 SYSTEM-OF-SYSTEMS MODELING IN STK

Every satellite system is expected to operate within a larger system of systems. For a satellite system to achieve its mission objectives, it needs input from ground systems. Today's technological advancements require simulation tools to model systems of systems in order to fully understand system-level interdependencies and operations. In STK, you can configure and analyze each subsystem and system and then save entire system configurations. These system configurations can then be loaded into larger, more complex scenarios so that system-to-system interactions can be modeled.

There are endless mission concepts that STK can be used to analyze. This section will cover some common system-of-system applications that STK is used for.

2.4.1 Constellation Design and Network Modeling

As launch vehicle technology advances, launches are becoming easier and more affordable. Constellations of satellites are being designed and operated by governments and commercial space. STK makes constellation design and analysis simple by enabling users to specify the following walker constellation configurations.

- Delta configurations have orbit planes distributed evenly over a span of 360 degrees in right ascension.
- Star configurations have orbit planes distributed evenly over a span of 180 degrees in right ascension.
- Custom configurations enable the user to define the span of right ascension to distribute orbit planes over.

If these configurations do not meet the engineer's needs, they can create or load in any other satellite to add to a constellation object. Constellation objects are used to condense analysis of many objects into the analysis of one object.

Example

To evaluate the communications between a constellation of LEO satellites with a ground network, you can create two constellations and evaluate the chain link between them. First, a constellation object is defined containing all the LEO

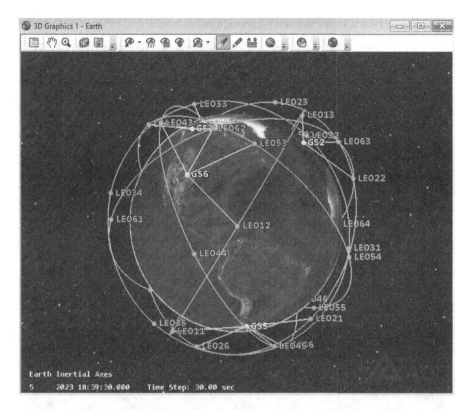

FIGURE 2.14 A constellation of satellites communicates with a ground network.

satellites. Next, a constellation object is created that contains only ground station objects. A chain is used to link the constellation of satellites to the constellation of ground stations and evaluated. Analysts can evaluate which satellites can communicate with which ground stations at any given time during the simulation. In STK, active communications links are visible as shown in Figure 2.14.

By analyzing direct satellite-to-ground links, analysts can determine if multi-hop links are necessary based on constellation geometry and ground network availability. If necessary, a satellite with no ground station access within a constellation can be designed to communicate with another satellite near it to relay data through the neighbor satellite and to the ground. This type of network routing can be configured and analyzed using STK. A user can load in network routing files to define priorities for inter-asset link building. If keeping a minimum number of "hops" from asset to asset is the top priority, it can be assigned as such. If completing a link from an asset to the ground as quickly as possible is the priority, it can be set. The result is a constellation interacting with itself and other assets in STK in a customizable way.

It is worth noting that STK can handle many (100,000+) satellite objects using a satellite collection object. The satellite collection reads in a database of satellites or a walker configuration into one object. This object can contain subsets of satellites that can be edited and picked out of the entire collection, but individual satellite properties cannot be. This object type is useful for high-level analysis and visualization at a system-of-systems level.

2.4.2 Rendezvous and Proximity Operations

STK can be used to model rendezvous and proximity operations (RPO). RPO involves two satellite systems of interest: one satellite must catch up with another in order to dock with it or collect imagery. The mission planning and analysis required to execute operations like this is complex. Complicated sequences must be executed at the correct times to ensure the mission is completed without unintended collision. STK has built-in RPO sequences that can be combined and run using Astrogator to simplify this process. In an RPO scenario, there is a target satellite and a chase satellite. The user can specify where the chase satellite should approach from, whether it should perch at some distance or scan the target object, and how long the sequence can take. The RPO mission planner uses these input variables and constraints to build dynamic scripts that solve for the necessary behavior of the chase vehicle.

To analyze RPO sequences, distances and velocities are often evaluated in the satellite's RIC (radial, in-track, cross-track) frame. Graphs of RIC position from one satellite to another are useful for evaluating relative position and motion. Figure 2.15 demonstrates the RIC values of a capsule as it docks with the ISS (International Space Station).

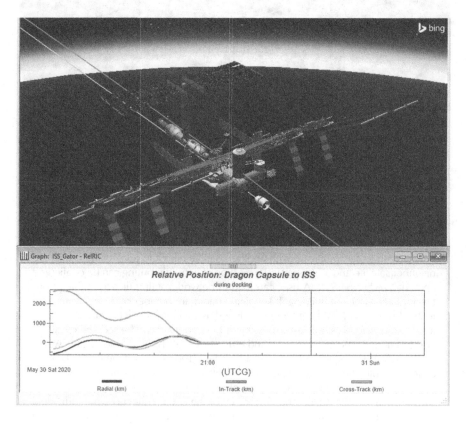

FIGURE 2.15   RIC values trend towards zero during docking.

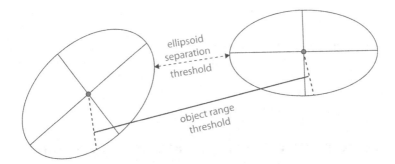

FIGURE 2.16   Range between threat volumes of orbiting bodies.

2.4.3 Collision Avoidance

There are thousands of Earth-orbiting objects in space and the number is constantly growing. Although space is vast and objects do not often cross each other's paths unintentionally, collisions can and do happen. They can create catastrophic volumes of debris, some of it trackable, some of it not trackable. To avoid satellite collisions, each satellite that is launched should be able to perform collision avoidance maneuvers in case another object's trajectory crosses its path.

STK can be used to analyze a system of interest and its position relative to other known objects in space. This analysis can be used to detect close approaches with all satellites in STK's public database or privately loaded database files. A more advanced conjunction analysis tool (CAT) is also available. This tool incorporates error ellipses to a primary list of objects and a secondary list. The primary list is composed of satellites of interest. The secondary list is composed of objects on potential collision courses. Advanced CAT-assigned ellipsoidal threat volumes to primary and secondary objects and a warning is issued if the range falls below a user-selected threshold. If the range between the threat volumes falls to zero, a collision occurs. A two dimensional depiction of satellite threat volumes is shown in Figure 2.16.

2.4.4 Behavior Execution Engine

With rising interest and requirements driving the move towards model-based systems engineering (MBSE), space systems are being designed and analyzed in a new way. They can now be described using modeling languages like SysML (systems modeling language). SysML diagrams may describe properties of a mode, how the system transitions from state to state, or how a sequence is executed. When combined, these diagrams are a complete view of the system and how it behaves. Abstracting up a level, it is important to understand how multiple systems interact with each other and what unforeseen behaviors may emerge as they interact. The environment that the systems interact within is also important as it may affect how each system behaves. Modeling a system in its physical environment requires a digital twin; a digital twin is a simulated copy of the physical system and can be described by models such as SysML diagrams.

Ansys's Behavior Execution Engine (BEE) is a model execution and integration engine for executable architectures. It integrates MBSE artifacts (SysML) with environmental analysis tools (STK) to create a time-synchronized, event-based architecture to represent a system. BEE is used to analyze how disparate systems behave and interact in a common environment such as STK. It can also be used to assess how well MBSE designs perform relative to design reference mission (DRM) constraints and requirements. Additionally, BEE's relationship to SysML allows for straightforward identification of the root causes when a requirement is not met at any phase of the engineering life cycle.

2.5   AN OVERVIEW OF ORBIT DETERMINATION TOOL KIT

From mission design through operations, GNC engineers need to understand their satellite's orbit. The dynamical models used to determine a satellite's orbit are exactly that, models. Though STK's models perform well, OD is necessary to ensure a satellite is robust enough to handle uncertainty and maneuvers for long-term operations. OD combines dynamical models (or predictions) and measurement data using a filter. Models are not perfect matches to reality and neither are measurement data, since measurements are biased and noisy. The best estimate of a satellite's orbit requires incorporating measurement data with a dynamical model through a filter.

Ansys's ODTK provides highly accurate orbit estimates. It has been used in every phase of the life cycle from design to operations on high-profile missions. It combines the powerful physics-based dynamical models available in STK with more than 100 available measurement types, including the following:

- Ground-based, relay-based, and space-based tracking
- Multi-GNSS support including pseudo-range and carrier phase
- Optical navigation
- Geolocation observations/Time difference of arrival/Frequency difference of arrival
- Accelerometer

To use ODTK, a user begins by inserting the satellite(s) of interest into the scenario as well as other assets that can provide measurement data about the satellite(s) of interest. These assets could be ground stations with trackers, GNSS satellites, or custom satellites. Measurements from these assets can be brought into ODTK from physical data (telemetry of an operating satellite) or by using a simulator within ODTK. These measurements are then filtered with a configurable dynamics model to estimate the orbit and its covariance (which can be ingested into STK). Many plots and reports are available in ODTK to quantify the OD and determine confidence in the OD solution.

It is simple to integrate with ODTK through many programming languages. This makes it simple to automate ODTK scenario generation, pull ephemeris and covariance data from ODTK into external tools for operational purposes, and build utilities that simplify commonly executed tasks in ODTK.

3 The System Effectiveness Analysis Simulation (SEAS)

Eric Frisco

3.1 INTRODUCTION

3.1.1 SEAS OVERVIEW

The System Effective Analysis Simulation (SEAS) is a constructive modeling, simulation, and analysis tool that enables mission-level military utility analysis. SEAS offers a powerful agent-based modeling environment that allows the analyst to simulate the complex, adaptive interactions of opposing military forces in a physics-based battlespace. Agents execute programmable behavioral and decision-making rules based on battlespace perception. The interaction of the agents with each other and their environment results in warfighting outcomes. The ability to represent networked military forces reacting and adapting to perception-based scenario dynamics makes SEAS ideally suited for exploring new warfighting capabilities, in particular, those provided by military space systems.

As an accepted analysis tool part of the Air Force Standard Analysis Toolkit, SEAS has been used in support of numerous Air Force and Department of Defense (DoD) studies. Figure 3.1 shows the SEAS application running a sample satellite communications (SATCOM) study scenario.

3.1.2 SUPPORTED COMPUTER PLATFORMS

The current version of SEAS as of this writing is 4.0, which marks the first major upgrade to SEAS in over three years and brings a modern user interface and several new features for streamlined modeling, simulation, and analysis. As a cross-platform application, SEAS is supported on standard MS-Windows systems, macOS, and various Linux distributions. No special hardware or software is required for the baseline configuration. Users running large simulations with many thousands of agents executing complex behavioral logic will benefit from having additional CPU/GPU horsepower and memory.

3.1.3 SOFTWARE COMPONENTS

SEAS is a single application written in C++. Previous versions of SEAS were integrated with the Eclipse Integrated Development Environment (IDE), but the model code editing and error-checking features of Eclipse are now built into the SEAS user interface. SEAS is released with an integrated help file system, user manual, and several example model files. Also included in the release are files for planet imagery, icons, and algorithms for satellite propagation.

DOI: 10.1201/9781003321811-5

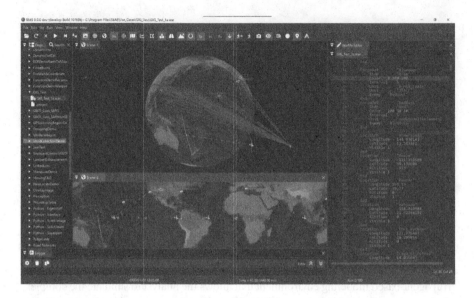

FIGURE 3.1   The SEAS application running a sample SATCOM scenario.

3.1.4   Release Policy

The Space System Integration Office (SSIO) at U.S. Space Force, Space Systems Command, manages the development and distribution of SEAS. Requests for the software can be issued to SEAS government leads. Contact details can be found on the SEAS website, at https://www.teamseas.com.

3.1.5   Deployment

SEAS is deployed from an installer, which is separately targeted for each computer architecture (Windows, macOS, Linux). The complete installation of the software and associated data files is approximately 4 GB. Most deployments of SEAS are on stand-alone computer or local area networks, but SEAS has been approved by the Air Force Network Integration Center (AFNIC) for installation on NIPRNet and SIPRNet systems. SEAS also has been approved for use on JWICS and other networks, by the appropriate network-approving authorities.

3.2   DEVELOPMENT

3.2.1   Development History

The SEAS program began in the early 1990s to support space force modernization. In order to establish investment priorities for future space system acquisition programs, the Air Force needed a way to evaluate the potential effectiveness of alternative space system concepts in the context of warfighting scenarios. A detailed review of existing scenario-based mission and campaign tools identified significant shortcomings for modeling space services. SEAS was created to address these deficiencies and enable broad exploration of space system architecture effects on major combat operations. Initial development of SEAS focused on the ability to explicitly model the sensor-to-shooter link and capture the non-linear behavioral impact of command, control, communications, computers, intelligence, surveillance, and reconnaissance (C4ISR) on spatial/temporal maneuver and attrition of terrestrial forces.

3.2.2 DESIGN PHILOSOPHY

SEAS was built to analytically assess the link between space systems performance and combat outcomes. Development of SEAS was shaped by a Government Accounting Office (GAO) report [1] identifying the need for such a tool and RAND Project Air Force studies [2] suggesting an agent-based simulation framework as the best approach. The design philosophy focused on (1) explicitly capturing the probability graph linking space services to how the joint force was fought; (2) explicitly modeling the observe, orient, decide, and act (OODA) loops embedded in the agent hierarchies instantiating the probability graph; and (3) capturing the probability density functions for Measures of Performance (MOPs), via Design of Experiment (DoE) and Monte Carlo simulation, allowing explicit delineation of the pathways between space system performance and combat outcome distributions. The last point is extremely important as it allows the assessment of risk associated with specific combat outcomes against space service performance.

3.2.3 KEY FEATURES

The SEAS simulation engine has several features that make it a powerful, flexible, scalable, and modern tool capable of modeling large-scale conflict scenarios. The following sections describe some of the key features of SEAS that have led to its widespread use.

3.2.3.1 Agent-Based Modeling Environment

SEAS offers a powerful agent-based modeling environment that allows the analyst to simulate the complex, adaptive interactions of opposing military forces in a physics-based battlespace. Agents (units, vehicles, planes, and satellites) execute programmable behavioral and decision-making rules based on battlespace perception. The interaction of the agents with each other and their environment results in warfighting outcomes.

3.2.3.2 Integrated Development Environment and Debugger

SEAS includes an integrated development environment that provides a single interface for developing, managing, debugging, and running SEAS model files. Tactical Programming Language (TPL) syntax is parsed in real-time and syntax errors are highlighted.

3.2.3.3 Flexibility in the Hands of the Analyst

The programmable and constructive nature of SEAS allows the analyst to create models with varying degrees of fidelity, resolution, sophistication, and complexity.

While SEAS is typically used for scenario-based military operations research involving aircraft, satellites, ground vehicles, communications systems, weapons and sensors, its flexibility enables it to be used to analyze a broad range of complex adaptive systems, not just military scenarios.

3.2.3.4   Analytical Graphics and Visualizations

SEAS produces graphic visualizations as a simulation is running. The user interface provides several different view options for displaying map features, sensors fields, satellite orbits, and output plots as results are generated. The analyst can also completely control the output display programmatically with TPL. This powerful feature enables improved debugging and allows the analyst to convey specific model details and results.

3.2.3.5   Optimized Performance

Most agent-based simulations have performance limitations and do not scale well since the number of potential agent interactions increases exponentially as the number of agents increases. SEAS tackles this problem with a rich set of advanced algorithms and techniques for efficient simulation and optimization. The result is a simulation system that offers high performance and scalability, with no special hardware requirements.

3.2.4   DEVELOPMENT TEMPO

The development of SEAS has been driven by study needs for over two and a half decades. One of the original goals for SEAS was to support "quick reaction analysis" and the ability to perform studies in the timeframe of weeks rather than months as was common with other tools. This led to a study-driven development cycle analogous to "just in time" manufacturing. This approach proved fast and efficient for quick reaction analysis and is still being used today. SEAS software releases are based on feature sets that have been rigorously tested performing studies and analyses.

3.3   MODELING FRAMEWORK

SEAS can be described as a space and terrestrial "sandbox" for agent-based simulation of conflict. The SEAS modeling framework consists of three primary components: (1) agents, (2) devices, and (3) environment. Agents, represented in SEAS as units, vehicles, planes, and satellites, span three domains: physical, information, and cognitive. The physical domain essentially defines the attributes of the agent in terms of simple performance parameters and their equipment, or devices (communications gear, sensors, weapons). The information domain represents the ability of an agent to have "memory" and battlespace perception by storing target detections, commands or instructions, and other kinds of information. The cognitive domain represents the "brain" of an agent defined by user-programmed or default behavioral logic. During the running of a simulation, agents execute OODA loops constructed by the user. As the agents interact through their devices with each other and their environment, warfighting outcomes emerge. Figure 3.2 depicts the conceptual representation of a SEAS agent.

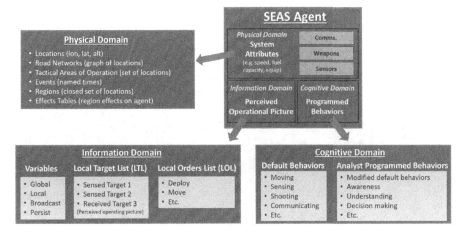

FIGURE 3.2 A conceptual representation of a SEAS agent.

3.3.1 War Files

The input to the SEAS simulation engine is referred to as a "war file." A war file is a text file (with a. war file extension) that defines agent hierarchies, their command structure and interconnection web of missions, the environment, devices owned, and doctrine-driven orders for moving in the environment and using devices (communications gear, sensors, and weapons) toward mission goals. Fundamentally, the war file is an instantiation of a dynamic probabilistic graph with its explicit conditional independence functions. War file syntax is referred to as the TPL, which is discussed in more detail later in this chapter. The SEAS object hierarchy is represented in Figure 3.3.

3.3.2 Agents

There are four agent types in SEAS: (1) unit, (2) vehicle, (3) plane, and (4) satellite. Each agent type in SEAS is defined by a set of attributes that establish its physical and information domain properties and an orders block of code that represents its cognitive domain properties with behavioral logic programmed in TPL. Each agent type also has default behaviors. The primary difference among the agent types is how movement and resources are managed by the simulation engine. The SEAS user has the flexibility to select the appropriate agent type for its desired use or functionality in the simulation.

3.3.2.1 Units

Units are the most general agent type and can be used to represent physical or nonphysical entities in the simulation model. Units are child objects to a force, and can have their own child objects, including other units, vehicles, planes, and/or satellites. Units modeled in SEAS typically map directly to a military unit, such as a "distributed common ground station" (DCGS), or a force hierarchy element, such as

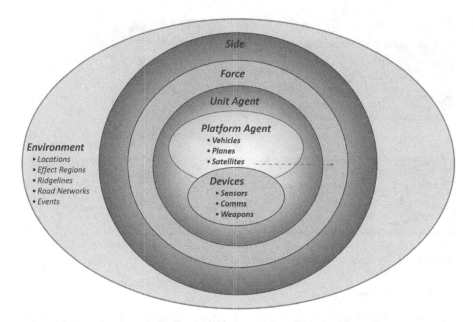

FIGURE 3.3   The SEAS object hierarchy.

the "18th Space Defense Squadron." In addition to constructing hierarchies of units representing a military force structure, units are also typically employed in a model to manage various aspects of the simulation, such as view displays, camera controls, file operations, intermediate output, and more.

3.3.2.2 Vehicles

Vehicles are objects in SEAS that have a child relationship to unit agents or other vehicle agents. This agent type also provides great flexibility in terms of the user's ability to control its movement and manage its resources, such as fuel consumption. Because of this modeling flexibility, vehicle agents in SEAS typically represent ground platforms, such as tanks, as well as air platforms, such as unmanned aerial systems (UAS), fighters, bombers, etc. Depending on the fidelity of the model, vehicles are sometimes used to represent individual soldiers or noncombatant personnel.

3.3.2.3 Planes

Planes are objects in SEAS that have a child relationship to unit agents. This agent type is typically used to model aircraft by analyzing simple behaviors, such as point-to-point sorties and basic loitering. Advanced users typically use vehicle agents for more complex behavioral modeling of piloted and unmanned air platforms.

3.3.2.4 Satellites

Satellites are objects in SEAS that have a child relationship to a Force or a unit. This agent type almost always is used, as the name implies, to model satellites that follow an Earth-centered orbit path. There are three methods for defining satellite orbits: (1) the SEAS engine can use numerical methods to create orbits from a set of standard orbital parameters; (2) two-line element sets, or TLEs, can be propagated using the built-in open-source SGP4 propagator or, optionally, using the SGP4 propagator in the Standardized Astrodynamic Algorithm Library (SAAL); and (3) SEAS can read pre-calculated orbit positions, in terms of longitude, latitude, and altitude, as a function of time.

3.3.3 Devices

There are three device types in SEAS: (1) sensors, (2) weapons, and (3) communications gear. Devices represent the equipment that a military platform (defined as a SEAS agent) possesses and uses to interact with other agents and its environment. For example, an F-35 aircraft in SEAS, which typically would be modeled as a vehicle or plane agent, might have multiple devices such as radars, electro-optical systems, ground attack munitions, air-to-air missiles, UHF SATCOM, and line-of-sight comm antenna suites. There is no limit in SEAS to the number of devices associated with an agent.

3.3.3.1 Sensors

Sensors are device objects in SEAS that can be owned by any agent type and are a way through which agents interact with each other and the environment. This device type has basic geometric constraints that define a field of regard and is used to "detect" enemy agent locations and velocities, missile launches, weapons fire, communications activity, and other active sensors. Detections are probabilistic based on user-defined entries in a Probability of Detection (Pd) table that provides a per-minute detection probability for a given sensor and target pairing. Sensor definition attributes in SEAS determine whether a sensor is active or passive and whether targets can be sensed when they are stationary, moving, flying, grounded, shooting, and/or radiating. Sensors can measure errored position and velocity, perform automatic battle damage assessment (BDA), and provide contributions to the Probability of Identification (PId).

3.3.3.2 Weapons

Weapons are device objects in SEAS that can be owned by any agent type and are a way through which agents interact with each other. This device type has basic performance parameters (speed, range, firing rate, etc.) and is used to damage or kill opposing force agents. There are two kinds of weapons in SEAS: (1) direct fire munitions and (2) missiles. If a munition from a weapons fire hits its intended target within the defined simulation time step, it is considered a direct fire weapon, and the munition "fly out" is not modeled. For missile-type weapons, the SEAS engine will "fly out" the munition along a linear or ballistic trajectory, depending on the settings in the weapon definition. Hits, damage, and kills are probabilistic based on user-defined entries in a Probability of Kill (Pk) table for given weapon target pairings.

TABLE 3.1

Environmental Region Effects in SEAS

Effect Identifier	Applies To...	Description
Speed	Units Vehicles Planes Supply agents	Multiplies the speed of an agent by the factor if the agent's current time step of movement passes through the region
Pk	Weapons	Multiplies the Pk of a weapon/target pair by the factor if the shot passes through the region
Pd	Sensors	Multiplies the Pd of a sensor/target pair by the factor if the sighting passes through the region
Reliability	Comm gear	Multiplies the reliability of a transmitting comm device by the factor if the message passes through the region
Range	Weapons Sensors Comm gear	Multiplies the range of a weapon, sensor or transmitting comm. device, by the factor if the shot/sighting/message must pass through the region
CEP	Weapons	Multiplies the CEP of a weapon by the factor if the weapon passes through the region

3.3.3.3 Communications Gear (Comms)

Comms are device objects in SEAS that can be owned by any agent type and are a way through which agents interact with each other. This device type has basic performance parameters and is used to pass messages back and forth between agents. SEAS has a first-order communications model that addresses the physic of closing the link (range, etc.), capacity of the link (bandwidth, etc.), message type (target detections, orders, broadcast variables), message size (bits), message class (user-defined access groups), message prioritization, routing strategies, and antenna beam patterns.

3.3.4 ENVIRONMENT

Environment objects in SEAS are used to define positions, paths, regions, ridgelines, road networks, and terrain. Agents can be deployed to locations and move along defined paths, road networks, or terrain. Both agents and devices can be impacted by environmental objects, such as a region, which can be used to model a number of effects without the computational burden of modeling weather, terrain, or buildings explicitly. Table 3.1 describes the SEAS region "effect table," that is available to achieved various effects.

3.4 TACTICAL PROGRAMMING LANGUAGE

The TPL is an interpreted language unique to SEAS that both defines and models the "brains" of an agent, telling it how to interact within its own command hierarchy and how to react and adapt to its environment and opposing forces. Each agent in SEAS contains an orders block where the behavioral logic is coded in TPL.

3.4.1 PROGRAMMING CONSTRUCTS

The programming constructs in TPL are similar to other programming languages, so having coding experience before using SEAS is beneficial. Table 3.2 lists the available programming constructs in TPL. Syntax, detailed descriptions, and code examples can be found in the SEAS help and user manual installed with SEAS.

3.4.2 DEFAULT BEHAVIORS

While TPL allows the SEAS user to control many aspects of a simulation and the behavior of the agents being modeled, it is important to note that the SEAS engine has several built-in, or "default," behaviors for each agent type (units, vehicles, planes, satellites). While advanced users tend to override many of the default behaviors, the default behaviors enable models to be constructed very quickly. This is especially useful for users with limited coding expertise. Default behaviors exist for movement, sensing, communications, weapons fire, resource consumption, and agent lifecycle status.

3.4.3 TIME-STEP EXECUTION

SEAS is a dynamic time-step simulation with a default time step of one minute. When SEAS runs a war file, the engine creates instances of all agents, setting attributes for each to values provided in the agent definition or using defaults if initial values are not specified. The order that agents appear in the war file hierarchy determines the order in which they receive the execution pointer at each time step. On a

---

TABLE 3.2

Programming Constructs Available in SEAS TPL

Programming Construct	Description
Comments	Code that is not executed; used for documentation
Variables and Data Types	Four classes of variable scope: global, local, broadcast, and persist. Eleven variable types: real numbers, location variables, strings, nodes/linked lists, road networks/road nodes, pointers, arrays, 3D vectors, state vectors, colors, maps
Operators	Basic math operators, vector and scalar operators, logical operators.
Conditional Statements	Logic control using If, Else, ElseIf, statements
Loops	Logic control using Loop and While statements
Subroutines	Named sections of code that can be called by other code
Functions	Built-in or user-defined code for performing various operations and returning a value given an input
Commands	Built-in logic and operations for various functionality: movement, communications, satellite maneuvering, file operations, etc.
Properties	Read, write, and/or read-writable values associated with agent, device, environment, and simulation objects

macro scale, it appears that agents act in parallel, but the simulation engine executes each agent's instructions sequentially. SEAS uses a serial process of freezing simulated time and allowing each agent a "turn" to execute its TPL code. This time-step method allows all agents to work off a common picture of the "state the world" (the environment, state, and positions of platforms and devices, etc.) at a frozen moment in time; the end of the last time step. SEAS executes each line of TPL inside each agent's orders block until it reaches a statement that causes that agent to surrender the execution pointer to the next agent, effectively ending the agent's turn to make decisions and take action until the next time step. All the commands that agent had queued that could alter the state of agents or the environment will execute during the adjudication phase. After all the agents have had their "turn," the simulation time is advanced one time step and all fires, detections, message transmissions, and trajectories are updated based on the physics and statistical models inside the SEAS engine.

3.4.4 PYTHON INTEGRATION

A powerful feature of SEAS is the ability to extend the capabilities of the TPL with calls to Python, one of the most popular and widely used programming languages today. The SEAS engine has an embedded Python interpreter to allow native Python code to be called from TPL code. This gives the SEAS user convenient access to a host of open-source code libraries that contain comprehensive mathematical functions, optimization algorithms, data plotting routines, and more. As of the time of this writing, an installer is available for SEAS that includes Python 3.9 with the libraries listed in Table 3.3.

TABLE 3.3
Python Libraries Included with SEAS

Python Library	Source
Basemap 1.2.2	https://github.com/matplotlib/basemap
FLANN 1.9.1	https://github.com/mariusmuja/flann
GDAL 3.2.1	https://pypi.python.org/pypi/GDAL
Kiwisolver 1.3.1	https://pypi.org/project/kiwisolver/
Matplotlib 3.3.4	https://matplotlib.org/
NetCDF4 1.5.5.1	https://github.com/Unidata/netcdf4-python
NumPy 1.19.5 + Intel Math Kernel Library	http://www.numpy.org/
Pandas 1.2.1	https://pandas.pydata.org/
Pillow 8.1.0	https://python-pillow.org/
PyKalman 0.9.5	https://pykalman.github.io/
PyWavelets 1.1.1	https://pywavelets.readthedocs.io/
SciPy 1.6.0	https://www.scipy.org/
Scikit_image 0.18.1	http://scikit-image.org/
Scikit_learn 0.24.1	http://scikit-learn.org/

3.5　APPLICATIONS OF SEAS

SEAS has supported numerous space system architecture studies, wargames, Analysis of Alternatives (AoAs), and various military utility analyses over the years. A typical example, presented here, is the Presidential Directive Memorandum III study (PDM III) conducted for the Office of the Assistant Secretary of Defense for Networks and Information Integration (OASD NII). This unclassified, multi-domain, theater-level study analyzed airborne and ground wide-area network options to meet local-area network connectivity requirements with existing and planned airborne, ground, and SATCOM (including commercial) platforms in a large, multi-service scenario. As is often the case with larger studies, SEAS was loosely integrated with two other simulation tools (OPNET and CAST) as indicated in Figure 3.4. The study was conducted over a ten-month period and coordinated with the joint staff, OSD Program Assessment & Evaluation (PA&E), and the National Reconnaissance Office (NRO).

The scenario involved a joint naval, marine, and army operation to regain control of the Main East West Road as indicated in Figure 3.5. The terrain was an eluvial floodplain surrounded by rugged mountainous terrain. Land-based avenues of approach were few and constricted. The attack is spearheaded by marine ship to objective maneuver (MV-22 Ospreys) followed by a simultaneous amphibious and army armored brigade assault from the south.

The time evolution of the scenario and the force structure options evaluated are indicated in Figure 3.6. The SEAS force hierarchy contained over 3,000 individual agents for this scenario.

Key measures of performance were blue casualties, vehicle loss exchange ratio, and time to secure the East West Road. The design of experiment called for 120 Monte Carlo

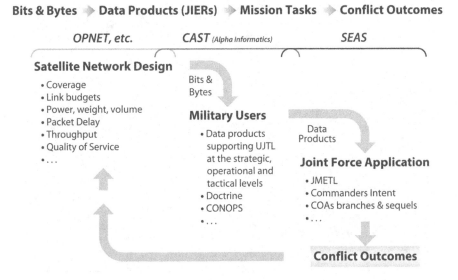

FIGURE 3.4　Study analysis flow across simulations.

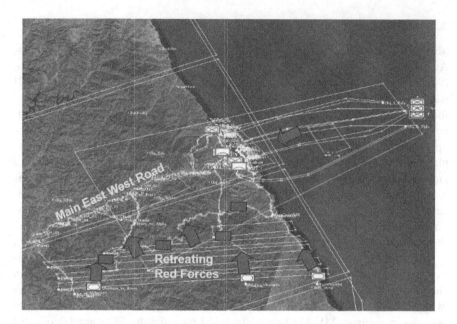

FIGURE 3.5   Joint force attack plan to seize East West Road.

runs for each force configuration. Distributions for each of these MOPs against each force structure are shown with violin plots in Figure 3.7.

These results are typical of the rollup measures produced for large studies. There are, of course, large amounts of intermediate data that allow tracing-specific mission and system effects through each battle realization.

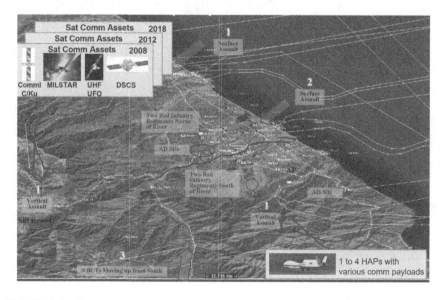

FIGURE 3.6   Scenario evolution and force structure studied.

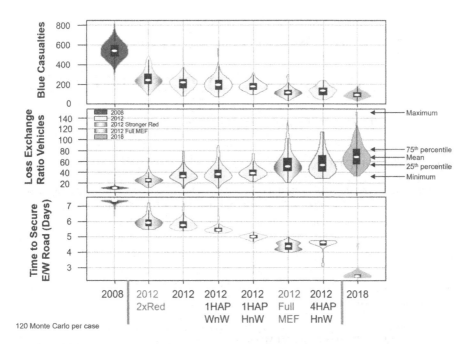

FIGURE 3.7   Measure of Performance (MOP) distributions vs. force structures studied.

3.6   SUMMARY

SEAS is a powerful, flexible, and robust modeling, simulation, and analysis tool that has been used for nearly three decades for exploratory and quick-reaction analysis in support of military space modernization and acquisitions. The ability of SEAS to model opposing military forces in an agent-based framework allows warfighting outcomes to emerge as agents interact with each other and make perception-based behavioral decisions. SEAS has proven to yield valuable insights for assessing the warfighter impact of space services on terrestrial conflicts.

REFERENCES

1. United States General Accounting Office (GAO), Defense Acquisitions, *Improvements Needed in Military Space Systems' Planning and Education*, Report to the Chairman, Subcommittee on Strategic Forces, Committee on Armed Services, and to the Honorable Robert C. Smith, U.S. Senate, May 2000.
2. D. Gonzales, L. Moore, C. Pernin, D. Matonik, and P. Dreyer, *Assessing the Value of Information Superiority for Ground Forces – Proof of Concept*, RAND Corporation, 2001.

4 The Satellite Orbit Analysis Program (SOAP)

David Y. Stodden and John M. Coggi

4.1 INTRODUCTION

The Satellite Orbit Analysis Program (SOAP) is a general-purpose software suite developed at The Aerospace Corporation which runs on standard desktop and laptop computers. Given the definition of initial conditions for point entities such as spacecraft, ground stations, and celestial objects, SOAP uses embedded propagation algorithms to predict past and future system behavior. The program presents a graphical user interface and animated displays of 3D scenes, plots, and alphanumeric results in a fashion that enables the construction and evaluation of analytical relationships involving complex systems and constraints. The setup and displays within SOAP are highly user configurable.

4.1.1 SUPPORTED COMPUTER PLATFORMS

Figure 4.0 depicts the SOAP screen running on macOS. The current version of SOAP as of this writing is 15.5.3. It is supported on MS-Windows 32/64-bit systems, macOS Intel/Apple Silicon Systems, and various Linux distributions. No special hardware or software is required for the baseline configuration. Some plug-ins, such as the Astro Standards Special Perturbations module, are subject to the terms of an NDA and must be obtained from the appropriate government agency. Users developing simulations involving large CAD files, or extensive terrain and imagery data, will benefit from having additional CPU/GPU horsepower and memory.

4.1.2 SOFTWARE COMPONENTS

SOAP is a monolithic application written in C and C++. Several helper applications are released with the package for various applications such as importing imagery and CAD models. Also released with SOAP are files for planet imagery, vector-based Earth maps, planetary shape/orientation data, collection of 3D CAD models, documentation/tutorials, and a large set of example cases. The program has an open architecture, meaning the users can add their own data resources to the mix. Several externally developed libraries are employed by SOAP are listed in Table 4.1.

Except for the Astro Standards, these are embedded in the SOAP executable.

4.1.3 RELEASE POLICY

The Satellite Orbit Analysis Program (SOAP) is developed at the Aerospace Corporation for the purpose of orbital and geospatial modeling and simulation (MS&A). It is primarily developed for internal corporate use and is not a deliverable.

 DOI: 10.1201/9781003321811-6

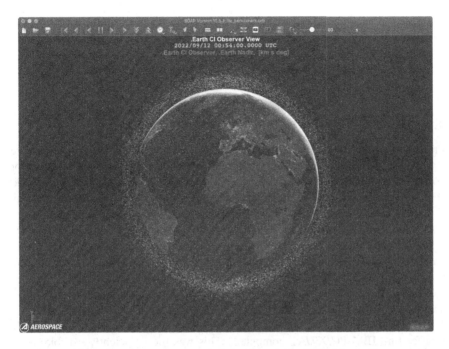

FIGURE 4.0   SOAP shows propagating 23,600 objects in the NORAD satellite catalog.

However, the policy on software reuse allows for the software to be provided for use on U.S. government programs. It is export-controlled in compliance with the Department of Commerce Export Arms Regulations (EAR-99) and is not available for purely commercial or educational use. Requests for the software can be issued to the through email, soap_af_poc@aero.org, and require an endorsement from U.S. government and approval from the Space Systems Command at Los Angeles Air Force Base.

4.1.4   DEPLOYMENT

SOAP is deployed in an installer, which is separately targeted and licensed for each computer architecture (Windows, macOS, Linux). The installer contains both programs and data, and each approaches a size of about 1 GB. Unique serial numbers

TABLE 4.1
External Software Components Used by SOAP

Library	Source
Qt Portable User Interface	https://www.qt.io/
JPL C-SPICE Version N0067	https://naif.jpl.nasa.gov/naif/
Astro Standards 8.4	USSF SPOC
libgeographic, libgeotiff, libjpeg, libpng, libssh2, libtiff, zlib	Online repositories
libavcodec 9.17	https://libav.org/

are issued to each end user and must be entered during the install process for the software to be fully functional. Licensing is per-user, so a user can install the software on multiple computers of the same architecture under their custody. An unlicensed copy of SOAP can act as a SOAP viewer, which can only read cases produced and encrypted by licensed users.

4.2 DEVELOPMENT HISTORY

Precursors to the SOAP program date back to the early 1980s. At the time, there was a desire to preview orbital and spacecraft geometry before submitting cases for costly thermal analysis. Several clients were developed for the Evans & Sutherland (E&S) PS-Series vector display terminals driven by VAX hosts. The Surface Display System (SUDS) program allowed 3D CAD models to be displayed after being assembled from geometric primitives. The Interactive Verification of Orbit Requirements (IVORY) program enabled models to be viewed from the position of Earth center and the Sun. The Define or Verify Ephemerides (DOVE) program allowed an orbit to be constructed and viewed from different perspectives. Because the E&S terminals did not have CPUs, the orbital calculations had to be computed on the VAX host and downloaded to the terminals using FORTRAN WRITE statements.

In 1988, the Enhanced Graphics Adapter (EGA) was deployed for the rapidly proliferating IBM PC/XT/AT computers. This was the sufficiently capable raster display available to the masses. The ability to perform orbital MS&A at the desktop instead of the data center and the cheaper cost of the new PCs drove the demand to develop a client PC platform. The first version of the software was produced at the end of 1988, and incorporated the combined features of SUDS, DOVE, and IVORY, and was thus christened "SOAP." It was coded in Lattice C, and a custom driver for the EGA graphics was developed in 8086 Assembly Language. SOAP was to remain an MS-DOS application until 1995 and was distributed on floppy disks. During this period, SOAP clients were also produced for Motorola 68K Macintosh platform, NeXT Step, and Solaris. The program was widely used to model the performance of the developing GPS constellation during the Operation Desert Storm era.

In 1995, Windows 95 was released with an OpenGL Application Programmers Interface (API) that was adopted for the SOAP program, so PC production was moved to MS-Windows. Versions were also deployed on the Motorola 68000/PPC-based Macintoshes, NeXT Step, and Solaris. Due to the need to run on multiple architectures, cross-platform user interface software was used, first "Zinc" and then "Qt," which remains in use to this day.

During this time, a significant portion SOAP development was performed under contract to the Jet Propulsion Laboratory (JPL) for modeling NASA deep space missions such as NEAR and Cassini. The publicly available JPL CSPICE ephemeris library was integrated for modeling of both terrestrial and interplanetary missions. SOAP remains current with the latest version of CSPICE.

During the new millennium, SOAP became increasingly powerful along with the underlying hardware. From its humble beginnings of running within a single 64 KB address space, SOAP grew to encompass the full 4 GB 32-bit address space and beyond with the advent of 64-bit computing. Instead of a single orbit back in the E&S days, the entire Air Force satellite catalog consisting of over 23,000 objects can now be propagated simultaneously while maintaining a 30+ fps framerate. SOAP has also embraced some of the innovations that have come with networking and the revolution in Geographic Information Systems (GIS). In addition to orbital analysis, the software is now being used for terrestrial applications such as Counter Unmanned Aerial Systems (CUAS) and optimal ground antenna terminal placement in obscured environments. The development team is readying themselves for the challenges that lie ahead with cloud computing, digital engineering, and web clients.

4.3   DESIGN PHILOSOPHY

4.3.1   VERSATILITY VERSUS EASE OF USE

With the constant addition of new features, the SOAP program architecture has become increasingly complex. As with most software, design decisions involve versatility versus ease of use. SOAP favors versatility, making it a tool for advanced users. The functionality goes beyond just listing the "answers" and encourages users to "think outside of the box" to ensure that the right questions are being asked. With patience and diligence, the casual user should be able to master SOAP, though some knowledge of astrodynamics and geospatial concepts is helpful. Users who become frustrated with SOAP can contact the support staff listed in the help menu for technical support, reporting issues, and information regarding training.

4.3.2   PERFORMANCE VERSUS GENERALITY

Algorithm generality versus efficiency is another key tradeoff. Again, SOAP favors versatility at the expense of performance. This makes SOAP slower than some custom applications in solving specific sets of problems. The power of SOAP comes into play when the user encounters problems whose solutions lie across domains. The increasingly sophisticated hardware available to the user community has allowed improvements to both generality and performance.

4.3.3   PORTABILITY

SOAP is designed to portably exchange scenario and data files across Windows, macOS, and Linux file systems. Relative paths are used for references to external files, and a "Pack and Go" feature can be used to bundle up all dependent files into a zip or executable archive.

4.3.4   DEVELOPMENT TEMPO

SOAP is deployed as an in-house application, and the development must be responsive to emerging problems in the programs that Aerospace supports. Keeping the software architecture balanced in the face of incremental user requirements and funding has been a major challenge. There has often been pressure exerted to implement features that cater to a small set of users. Features that do not pertain to a broad user base are generally resisted or placed in an external utility application. SOAP is not certified for operational use.

4.4 RUNTIME ENVIRONMENT

SOAP is run by double-clicking the application icon, invoking it on the command line or by double clicking an associated *Orbit Scenario* or ".orb" file. When invoked without a file, SOAP presents an untitled scenario with default views and settings. The appearance of this is shown in Figure 4.1 and is described below.

4.4.1 TITLE BAR

The topmost portion of the SOAP window is the *Title Bar*. It contains "SOAP" followed by a major, minor, and increment version number. Following this is the name of the current scenario, shown without the path. A name of "untitled.orb" indicates a new scenario that has not yet been saved. If an existing scenario is loaded, pressing <shift><f> will open the folder containing that file. Pressing <shift><s> will bring up the folder of example scenarios.

4.4.2 THE MENU BAR

The *Menu Bar* is located directly beneath the title bar and is the basis for the SOAP pull-down menu system. The Objects menu is of particular importance, as this is the primary mechanism for creating new scenario content.

4.4.3 THE TOOL BAR

The *Tool Bar* is sandwiched beneath the menu bar and above the SOAP display area. Buttons are provided as user shortcuts for performing common file, clock, and viewing operations. Each button is invoked by positioning the mouse cursor over it and clicking once. Pull-down options exist for some buttons.

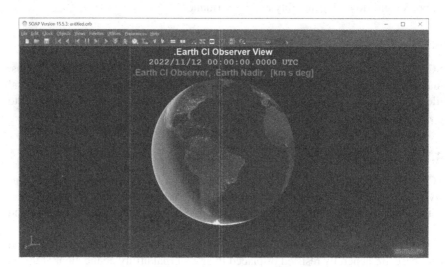

FIGURE 4.1   Capture of the MS-Windows SOAP start-up screen.

4.4.4 THE CLIENT DISPLAY RECTANGLE

The *Client Display Rectangle* is the portion of the window that contains the active scenario display. The initial screen displays a rotating Earth and a set of coordinate axes in a *World View* named the "Earth CI Observer View" as shown in Figure 4.1. The label "CI" is short for "Centered Inertial." In this view, the virtual observer is situated at a fictitious location in space—a rotatable *Platform* called the "Earth CI Observer". A virtual camera is pointed toward the Earth and oriented via a SOAP *coordinate system* called ".Earth Nadir."

SOAP views can be renamed and repurposed at will and the appearance of the client area can vary widely based on the scenario settings. Contained within the display area is a single *active palette*, which is a collection of one or more rectangular screen partitions. In SOAP, a palette can provide up to 16 rectangular panes within the single application window. The user can switch between multiple sets of palettes to select the active display. Each rectangular partition of a palette is designated as a *View Port* and contains a *View*. Views are assignable content that varies according to type. Views can contain 3D scenes (world views), plots (data table, parametric, XY, or color scales), 2D images, instantaneous data displays, reports, or text.

4.4.4.1 View Titles

Although the contents of the display vary widely according to view selection and state, there are common display features. When enabled, the *View Title* is by default displayed at the top center of each view. Each view must have a unique title, as this is the basis for selection and state information. Titles may contain up to 127 alphanumeric characters and are case-sensitive.

4.4.4.2 Clock Display and Object

When enabled, the *Clock Display* is shown in a selected format directly beneath in the selected view port underneath the view title. By default, the *Full Date* format is shown. The time is initialized to a user-definable date in the scenario called the *Epoch*. The default epoch for a new SOAP scenario is midnight, Coordinated Universal Time (UTC) of the host computer date. All times in SOAP are in UTC, but local offsets can be applied. SOAP obtains the value for the data from the host computer. The displayed clock format can be changed by double-clicking the time text in the active viewport. The simulation time is continuously incremented by another user-definable quantity called the *Time Step*. This is defined as the time increment between each animation frame. Tool bar and menu options allow the user to run the clock forward, backward, reset to epoch, and to increase or decrease the time step. The SOAP clock can be configured to run in real-time and synchronize with the host system time.

4.4.4.3 Status Line

When enabled, the *Status Line* is located directly below the simulation time. Its contents vary according to the view type. For the world view shown in Figure 4.1, it displays a *Platform ID* of a default SOAP platform object. In the context of a view, the platform defines the vantage point location. *Platforms* are simulation objects associated with user-defined or predicted locations. Next is a *Coordinate System* (CS) ID. These are objects used to establish the view camera direction and orientation. Last

are the currently defined *Distance, Time, and Angular* units. These settings control
the specification and display of input and output data. The properties of all status line
entities can be edited by double-clicking their text. The status line can be toggled on
and off using Shift-Z keys, or via the menu system.

4.4.4.4   Captions
At the bottom of the client area are a set of axes defining the Earth Centered Inertial
Coordinate System and the SOAP logo. Logos can be changed, and some commonly
used ones are included. Additionally, user-defined caption text can be inserted onto
optional header and footer lines, though these are not shown here. Optional headers
and footers can be displayed for classifications.

4.4.4.5   Current Scenario
When SOAP is launched without any scenario files, the default scenario is internally
generated. There are no satellites defined, but there are Earth, Moon, and Sun platforms
and a handful of platforms for the default views. There are a couple dozen pre-defined
Coordinate System objects. This scenario can be subsequently edited and saved by
the user to a named file. If users want to customize a SOAP start-up, they can specify
that a named scenario file be loaded as the default settings. Users often call up existing
scenarios using the file menu or dragging a file from the desktop into the SOAP main
window. A list of recently used files is provided under the file menu. The characteristics of SOAP scenarios are independent from the computer architecture being used
and the same scenario files can be shared across MS-Windows, macOS, and Linux.

4.5   OBJECT ARCHITECTURE

The essence of a SOAP scenario is composed of its list of defined objects and how
they interact. Objects can be imported from external data files or be defined interactively using the SOAP *Objects Menu*. A listing of the object types is presented in
Table 4.2.

The focus here is on foundational object types. The reader is referred to the *SOAP
User's Manual* for a complete description. Platform and coordinate systems are the
fundamental building blocks of a scenario, and nearly all other object definitions
depend on them.

4.5.1   PLATFORMS

In SOAP, *Platforms* represent moving points in space. The rules that govern how a
platform moves are embodied in internal algorithms called *Propagators*. Platforms
are categorized into different types, based on which propagation method is employed.
Once a platform is defined, it may have SOAP objects such as sensors, analyses, and
views attached to it. SOAP users can feasibly define and propagate up to 100,000
platforms simultaneously. The animation speed and program interaction may be
adversely affected if too many platforms are defined. Platforms are classified into
the subtypes listed in Table 4.3.

TABLE 4.2
SOAP Object Types

Object Type	Description
Platforms	Point entities with associated prediction models
Coordinate Systems	Defines matrices used for object pointing and orientation
Sensors	Platform geometric, optical, or radar payloads
Views	3D world, plot, image, data, report, or text displays
Palettes	Screen layouts consisting of 1–16 rectangular panes (view ports)
Analyses	Numeric and logical computational quantities
3D Models	Imported CAD geometry in SOAP "Surface-Based Model" format
Trajectories	Integration of relative motion between two platforms
Swathes	The path of a sensor field of view on a planetary body
RF Transmitters	Radio frequency transmitter payload and antenna
RF Receivers	Radio frequency receiver payload, antenna, and environment
Isochrones	Lines of position for signal time or frequency difference of arrival
Groups	Collections of platforms, analyses, or terrain/imagery grids
Actions	A scripted change to a SOAP object definition
Conditions	The criteria for performing an Action as referenced by the script
Delta-V	An instantaneous change imparted to an orbital platform
Data Grids	Data tables, contours, terrain/imagery tiles, parametric studies

4.5.1.1 Predefined Platforms

Predefined platforms are those whose orbital predictions are driven by external ephemeris files and thus cannot be changed in SOAP. Examples of planetary bodies are the Earth, the Moon, and the Sun. When JPL binary SPICE (.bsp) files are loaded, additional solar system planets, Barycenters, asteroids, and comets can be included, along with planetary moons. Except for Barycenters, which are point

TABLE 4.3
SOAP Platform Types

Platform Type	Description
Predefined	Planetary bodies, planet Barycenters, ephemeris files
Kepler	Ideal Keplerian propagation with secular J2 effects
NORAD	Versions of the SGP4/SDP4 propagators for use with TLEs
Low Thrust	A numerical integrator supporting perturbations
Airplane/Ship	Repeating great circles, figure 8s, and oval flight tracks
Air Route	Time-stamped latitude, longitude, altitude waypoints with loiter
Platform-Relative	A platform defined as an offset from an existing platform
Ground Station	A planet fixed latitude, longitude, altitude point
Transfer	Ballistic (Lambert) trajectory between launch and target platforms
Variable	A platform that changes identity based on a heuristic
RPO	Initial conditions for a rendezvous and proximity operations orbit

entities defining the common center of masses, massive bodies come with spherical, oblate spheroid, or triaxial ellipsoid shape models. Separate SPICE files are
used to model the shape, orientation, rotational rate, and gravitational parameters
of the planetary bodies. Predefined platforms can also be employed to model launch
vehicle or spacecraft ephemeris, for situations when the motion is too complex to be
modeled with simple two-body force models. An example is the Cassini planetary
tour, which took place during the years 1997–2004.

4.5.1.2   Kepler Platforms

Kepler platforms are used to model the motion of existing or proposed satellites in circular or elliptical orbits about a central body. They can also be used to model parabolic
and hyperbolic orbits. Inputs consist of a six-element state vector, an associated epoch,
a central body, and one of nine predefined *Initial Condition (IC) Types*. The Kepler
platform is often used with future studies, for which observation data are not available.
The propagator models perturbations resulting from the secular effects of the oblate
planet J2 gravity coefficient, but otherwise assumes ideal conditions. The value of J2
can be adjusted by editing the predefined platform acting as the orbit central body.

4.5.1.3   Low Thrust Platforms

Low Thrust Platforms share the same definition as the Kepler platform, but add a
Perturbation tab. This is a numerically integrating propagator that supports user-
defined thrust tables, atmospheric drag, solar radiation pressure, lunar, and solar
third body effects, and up to a 70x70 Earth geopotential model. The additional tab
has inputs for spacecraft mass, area, reflectivity, and a selection of models for atmospheric drag. The propagator is useful for force model studies.

4.5.1.4   NORAD Platforms

This platform type is defined by mean elements encoded into Two Line Element sets
(TLEs) as published in the NORAD space catalog. This propagator has a wealth of
observational data associated with it, and offers the user the advantage of simply
pasting data into SOAP rather than having to type them in. Various sources of TLEs
exist; a prominent one is: https://www.space-track.org/auth/login

The TLE data are associated with U.S. Space Force SGP4/SDP4 propagators.
SOAP includes these as dynamic libraries that come with the Astro standards version 8.4. Selected past versions of the propagators are available and are configured
by using the *Preferences*, *TLE Options* menu.

4.5.1.5   Air/Ship and Route Platforms

These platform types support the definition of moving vehicles on or above a planetary surface. The *Air/Ship Platform* is used to define repeating great circle, oval,
or figure-8 tracks. A *Route Platform* supports time-stamped latitude, longitude, and
altitude waypoints. Data can be imported from GPS receiver XML (.gpx) files for
playback into SOAP. The route platform supports loiters.

4.5.1.6   Ground Station Platforms

The *Ground Station Platform* allows the definition of a planet fixed point having a
given latitude, longitude, and altitude. On computers with Internet access, embedded

Google Maps can be used to search for and place ground coordinates of known locations.

4.5.1.7   Transfer Platforms

Given launch and target platforms, and a time of flight, the *Transfer Platform* can fit a ballistic trajectory between them using Lambert's method. An option also exists for a rectilinear transfer.

4.5.1.8   Platform-Relative Platforms

Given an existing platform, a platform-relative platform is defined at a perched location and an X, Y, Z, or theta, phi, magnitude offset based on a supplied SOAP coordinate system. Such platforms are often used at rotatable vantage points of their parent platform but are also defined for analytical purposes.

4.5.1.9   Variable Platforms

A *Variable Platform* can assume the identity of other existing SOAP platforms based on a heuristic. An example is to choose a satellite of minimum range from a spacecraft of interest. The variable platform will hop from location to location as the platform has the minimum range changes. This can be treated as a platform in the generic sense, and the range can be plotted, making an otherwise complicated problem simple.

4.5.1.10   Platforms Attributes and Notes

All platform types feature attributes and notes tabs. Attributes cover display properties such as icon type, color, and whether certain display attributes are enabled. The options available in this tab are shown in Figure 4.2.

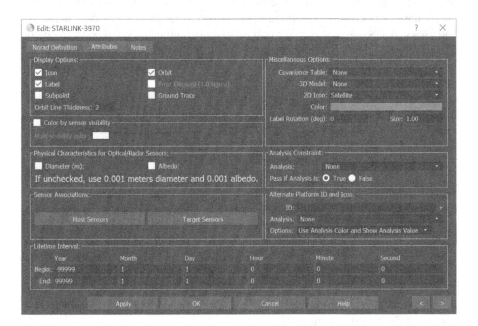

FIGURE 4.2   Platform attributes tab.

There are many options, not all of which will apply to every platform. By changing the state of the icon, label, subpoint, orbit, error ellipsoid, or ground trace checkboxes, the user is indirectly changing whether the platform appears in the view SET list in all SOAP world views. For example, enabling the orbit checkbox orbit will insert this into the active SET list in all world views. The orbit will appear in a world view if the orbit setting of the view is either SET or ON. Changing the state of a platform attribute here is a quick way of turning on or off that attribute in all world views at once.

Platform Notes are displayed as pop-up tooltips when the animation is off and the cursor is hovered over the platform icon (the tooltips option must be enabled in the preferences, universal panel). Rich text (i.e., bold, italic, or colored text) is supported via clipboard paste. Rich text, when saved to the orb file, is verbose due to HTML tags. To prevent SOAP from storing platform notes in rich text, select the save as plain text instead of rich text option.

4.5.2 COORDINATE SYSTEMS (CS)

Coordinate Systems are the unsung heroes of the SOAP scenario. They are seldom seen, having very few display attributes. Yet they work behind the scenes, directing the behavior of other objects. Many frequently used coordinate systems are pre-constructed for the user. However, these will only accomplish basic tasks. A mastery of coordinate systems is required for the SOAP virtuoso, and accomplished users should be able to build their own.

The SOAP CS definition differs from the commonly held notion of a coordinate system in one important way: the origin of each SOAP CS is usually not specified directly. Rather, the origin is inherited from the coordinates of a *Host Platform*, which are constantly being updated by SOAP. This binding is established with the definitions of the objects external to the CS, such as sensors, 3D models, analyses, and views. A CS is not effectively part of a simulation unless it is bound to one or more platforms in this fashion. The same CS definition can be bound to many simulation objects, eliminating the need for redundant CS specifications throughout the scenario.

The CS object definition allows references to other CS definitions. Such definitions can be chained together to satisfy the modeling requirements for many space systems. SOAP defines several CS subtypes that are listed in Table 4.4.

The basis CS is described in detail, and the reader is referred to the SOAP user's manual for a description of other types.

4.5.2.1 Basis CS

The *Basis CS* is the underlying model upon which all other SOAP CS types are built. All other types must at some point reference a basis CS to be completely defined. The basis CS panel allows the user to construct a three-dimensional coordinate frame by defining two non-parallel vectors directed outwards from each platform location. One vector defines the *Pointing Axis,* and the other is the *Reference Vector.* The vector cross-product of these two becomes the orienting axis, with the remaining unspecified axis constructed such that a right-handed system is formed. This concept is illustrated in Figure 4.3.

TABLE 4.4
SOAP Coordinate System Types

Object Type	Description
Basis	The fundamental CS type constructed from two vectors
Euler	Transforms parent CS by fixed or variable X, Y, Z rotations
Quaternion	Transforms parent CS by a quaternion-based rotation
Slew	Scheduled gimbals through a CS list using SLERP algorithm
C-Kernel	SPICE construct that interpolates a set of file-based rotations
Frame Kernel	SPICE construct used for planetary body orientation

All SOAP coordinate systems are right-handed. There is another important defining aspect of the basis coordinate system that transcends the mathematics. The pointing axis defines the direction of the assigned objects, such as sensors and views. The
projection of the reference vector into the plane perpendicular to the pointing axis
defines the reference orientation, or the 0° clock angle. When used in the context
of a SOAP worldview, this points toward the top of the screen. The SOAP basis CS
definition panel is shown in Figure 4.4.

Like all SOAP objects, a text field to define a unique ID is first and foremost.
This followed by the blocks to define the pointing axis and the reference vector. The

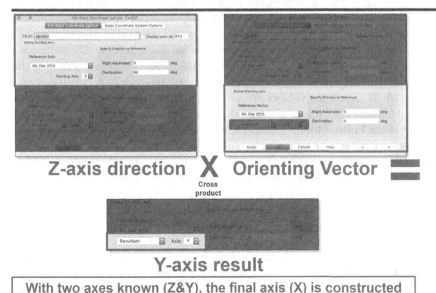

FIGURE 4.3   Constructing the default "EarthCI" coordinate system.

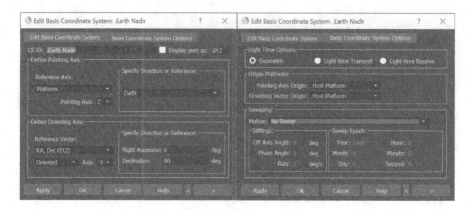

FIGURE 4.4   Definition of the basis coordinate system.

resultant of these two vectors is the orienting axis. The third, unspecified axis forms a right-handed system. It is important to know that the choice of X, Y, or Z axes does not affect the pointing and reference properties of the CS, although the letter designators must be different. However, letter designators can matter for dependent objects, such as 3D models, or axes-dependent references from higher-level coordinate systems.

4.5.2.2   Basis Coordinate System Pointing and Orientation Options

SOAP supports several vector types for defining directions for the pointing axis and orienting vector, and the user can draw from the same set of options for the pointing axis and reference vector. When a type is selected, the set of associated parameters corresponding to the selection appears next to it on the right side.

In the explanation of the options below, it is assumed that the CS under construction is bound to a platform with the identifier "K." (Binding occurs in the subsequent definition of Sensors, analyses, views, or 3D models as mentioned previously). In all cases, the vectors are assumed to originate at the position of K. Another identifier "B," refers to the geometric center of a planetary body of interest, such as the Earth, Moon, or Sun. This is selected from a pull-down menu next to the options that use it.

RA, Dec (J2000) option: Points in the direction of the specified right ascension and declination angles. The directions are based on an underlying J2000 coordinate frame. *Right Ascension* is the angle in the J2000 XY plane measured positive from the X-axis (J2000 vernal equinox) towards the Y-axis. *Declination* is the angle measured above or below the XY plane, positive in the direction of the celestial north pole.

RA, Dec (ECI) option: Similar to the option above, except a generic Earth Centered Inertial (ECI) frame is used in place of J2000. The definition of the ECI frame is based on that found in the "Earth CI/CR" planet orientation field defined within the preferences, universal panel as described in section. ECI can be set to *J2000, Precession* (true equator, mean equinox), *Nutation* (true equator, true equinox), or to an arbitrary orientation defined by a SPICE *Frame Kernel*.

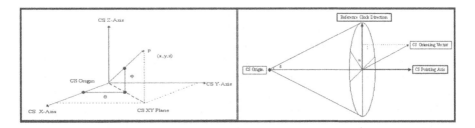

FIGURE 4.5   (Theta, phi), (clock, cone) angles.

Latitude/Longitude option: The direction of the vector is parallel to the vector from the center of the Earth to the specified longitude and latitude. The vector does not point to a (longitude, latitude) Earth location, unless K is at the center of the Earth. This option is like right ascension/declination pointing except that it is defined in an Earth-rotating frame rather than in an inertial frame. However, latitude is considered a geodetic angle for oblate spheroids.

Theta/Phi option: This option is a generalization of the RA, Dec options to encompass any underlying SOAP coordinate frames. The user specifies an existing SOAP coordinate system from a pull-down list and direction angles theta, and phi as offsets from the X-axis of this CS. Theta is the angle in the XY plane of reference CS and is measured positively from the X-axis towards the Y-axis (like right ascension in an ECI system). Usually, its domain is from 0–360°, though negative values can be used. Phi is the angle measured positive from the XY plane of the reference CS toward the Z-axis (like declination in the ECI system). Its domain is from -90–90°. The theta/phi option may be used to chain multiple coordinate systems together. Recursive references are trapped. The constructs defined in this section are illustrated in the left side of Figure 4.5.

Clock/Cone option: This option is an alternative way of specifying a direction with respect to existing SOAP CS. Unlike the theta/phi option, the two input angles are not based directly on the X, Y, or Z of the reference system. The clock/cone angles are illustrated on the right side of Figure 4.6.

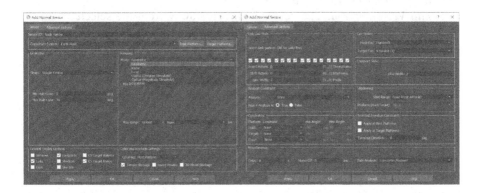

FIGURE 4.6   Definition of the normal sensor.

The *Clock Angle* (labeled α above) is defined as that between the *Reference Clock Direction* and a direction of interest. This angle is measured positive clockwise from the parent CS orienting vector. The domain of the clock angle varies from 0°–360°. The name comes from the clockwise orientation of this angle when looking along the coordinate system pointing axis from the coordinate system origin.

The *Cone Angle* (denoted by δ above) is defined as the angle between the pointing axis of the reference coordinate system and a vector directed along the surface of a virtual cone of revolution as shown above. The domain of the cone angle varies from 0°–180°, with 0° aligned directly along the pointing axis, and 180° pointing in the opposite direction. The clock/cone option may be used to chain multiple coordinate systems together.

Platform option: The vector points to the position of P, the platform specified as the reference platform. If K and P are at the same position, an error will result.

BCI Velocity Vector option: The vector points in the direction of positive Body Centered Inertial (BCI) relative velocity of K with respect to B. Such inertial frames can be based on the alignment based on J2000 or an epoch of date, as specified in the *non-Earth CI* planet orientation field. If the relative velocity with respect to B is zero, a definition error will be displayed.

BCR Velocity Vector option: The vector points in the direction of positive Body Centered Rotating (BCR) relative velocity of K with respect to B. This frame is useful for steering surface vehicles such as ships, airplanes, and planetary rovers. If the relative velocity with respect to B is zero, an error will result.

Radial Vector option: The vector points away from the center of B. This is precisely the opposite direction of the platform option above with respect to the chosen central body. If the origin platform is located at the center of the selected central body, a singular definition error will result.

Nadir Vector option: The vector is directed from K to the closet surface point on the selected central body B. The behavior is similar to that of the platform tracking option above, except that this formulation considers that the body may be an oblate spheroid or a tri-axial ellipsoid. If K is on the planet surface, the vector points inwards along the surface normal at K. This is a common option for looking at a central body from an orbiting spacecraft.

Zenith Vector option: The vector is directed away from the selected central body surface point closest to K, which is the opposite direction to the nadir option above. The behavior is similar to that of the radial tracking option, except that this formulation considers that the central body may be an oblate spheroid or a tri-axial ellipsoid. If K is on the planet surface, the vector points outwards along the surface normal at K. This is a common option for looking up from a ground site.

4.5.2.3   Basis Coordinate System Advanced Option Tab

This definition panel tab offers additional options, such as light time transmit and receive tracking modes, host platform overrides, and sweeping options for conical and pendulum scanning.

4.5.3 SENSORS

SOAP *Sensor* objects enable the SOAP user to construct *Field of View* (FOV) volumes and associated physical constraints. These are attached to ground or space-based platforms and are pointed and oriented using SOAP coordinate systems. The primary function of a sensor is that given a host platform, to provide visibility determination to one or more target platforms. In addition to whether a target may lie within the coverage volume, constraints such as planet occultation, 3D model blockage, and Sun illumination conditions can be applied. The three supported sensor types are normal, region, and link.

Normal sensors encompass a wide variety of geometric and behavioral options.

The defining panel for a normal sensor is shown in Figure 4.6.

The panel is partitioned into sections assigning coordinate system and platforms, ranging, display options, link patterns, shadowing, and analysis, elevation, and Sun constraints.

4.5.3.1 Sensor Coordinate System and Host, Target Platforms

A SOAP sensor must reference a coordinate system and at least one host platform to operate. An instance of the definition is placed on each host platform. Target platforms are optional, and if present will cause links to be drawn from each host when visibility conditions are met.

4.5.3.2 Sensor Geometry

Supported geometry shapes include sections of cones, spheres, and extruded polygons. Examples are shown in Figure 4.7. The first shape, simple conical, is also the one shown in the defining panel. The half-cone angle is measured outwards from the CS pointing direction or bore sight. There is an option not shown about that allows the half-cone angle to be defined as an Analysis, thus making it variable.

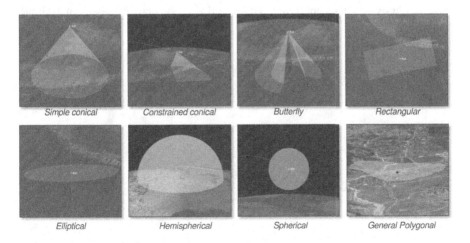

| Simple conical | Constrained conical | Butterfly | Rectangular |

| Elliptical | Hemispherical | Spherical | General Polygonal |

FIGURE 4.7   Sensor geometry shapes.

4.5.3.3   Sensor Ranging

Sensor range constraints are also defined and applied. SOAP can display these volumes in 3D space and can use them as criteria for visibility determination. SOAP supports several options for sensor ranging, which can determine the minimum and maximum range according to rules. These include *Geometry* (using the shapes described above), *Radar, Laser, Optical*, and *Data Table*. Geometric ranging includes specifications for minimum and maximum range with fixed and/or variable amounts. The screen in Figure 4.9 shows sensors being used in optical mode.

4.5.3.4   Sensor Display, Blockage Options, and Advanced Tab

These are shown in the bottom left screen of Figure 4.8 and allow the user to control which display options such as volumes, footprints, links, and shadows are displayed. Additional options regulate whether a sensor line of sight can be blocked by terrain or 3D models. The advanced tab defines additional constraints that can be applied to visibility determination and allows the user to configure additional rendering modes.

4.5.3.5   Region and Link Sensors

A SOAP *Region Sensor* is a single enclosed planetary geographical boundary extruded into space. This can be used to analytically detect when a platform is inside of its volume, e.g., determining if it is in a country's airspace. Databases for all the countries of the world are available as SOAP regions. Figure 4.8 depicts region sensors for California and CONUS.

The region sensor is also used to detect overlap between defined regions and satellite-based sensor ground coverage. In Figure 4.8, the conical sensor on "Sat 2" is shown to partially overlap California and CONUS, while its satellite remains outside of those airspaces. The rectangular sensor on "Sat 1" misses both regions entirely. SOAP also features an *Inscribed Area Sensor*, which is positioned at the center of a planetary body and does not have to be aligned with the body rotation. An example

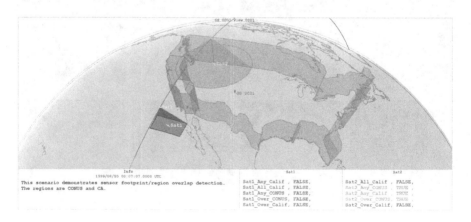

FIGURE 4.8   Region sensor shapes.

of this is a hemispherical volume directed toward the Sun to approximate the daylit portion of the body.

A *Link Sensor* is a lightweight version of the object that only considers planetary occultation and a defined timespan in drawing links between two specified platform. It is often used to represent connections defined by externally used software or conditions.

4.5.4 VIEWS AND PALETTES

Views are portals into the SOAP visual and analytical engine. They fall into categories such as 3D displays, plots, instantaneous data displays, and reports. A listing of the SOAP view types is provided in Table 4.5.

In SOAP, views are mapped to the screen in rectangular regions called *Ports*. Ports in turn are presented in *Palettes*, which are collections of one or more ports. Thus far, only world views and single-port palettes have been presented, but additional examples will reveal other view types and palette configurations.

4.5.5 ANALYSES

Analyses are SOAP Variables that return both logical and numeric results. There are presently 204 predefined analysis types, including quantities such as latitude, longitude, altitude, azimuth, elevation, and range. The reader is referred to the SOAP user's manual for the entire list. Analyses are classified into categories that mostly correspond to the type of units they return. A listing of these is provided in Table 4.6.

Highlighted are two of the more interesting analysis types, expressions, and constants. Editing screens for these two types are shown in Figure 4.9.

TABLE 4.5
SOAP View Types

View Type	Description
World	Renderings of space and planetary environments
XY Plot	Plots multiple SOAP analyses (variables) as function of time
Parametric Plot	X versus Y plots with time as the parameter
Data Grid Plot	Plots contents of 2-D meshes such as terrain heights or contours
Color Scale Plot	Used as a color legend for contours and some plots
Analysis Data	Instantaneous display of SOAP analysis results
Ephemeris Data	Instantaneous display of platform group ephemeris
Sensor Data	Instantaneous display of sensor host, target platform relations
Rise Set Report	Generated display for platform overflight passes
Statistical Report	Gathers time-based statistics for SOAP analyses
Encounter Report	Generates data on platform close approach interactions
Text View	Presents user-supplied text or HTML
Image View	Presents user-supplied images or time-stamped image decks

TABLE 4.6

SOAP Analysis Type Categories

Analysis Type	Description
Orbital Parameters	Quantities used to define orbits, such as semimajor axis
Cartesian Distance	Variables having distance units, such as range
Cartesian Velocity	Variables having velocity units, such as range rate
Angular Position	Variables having angular units, such as longitude
Angular Rates	Angular velocity terms, such as elevation rate
Velocity Direction	Angles describing terms such as heading or flight path
Rotations	Rotational (XYZ) angles, matrix, or quaternion components
RF Parameters	Radiofrequency payload terms, like gain or received power
Time Values	Variable having time units, such as elapsed time or platform age
Visibility Counts	Integral counts of conditions such as targets in view
Logical Results	True or false conditions, such as target in view
Constant	User-defined numeric values with defined units
Expression	Logical or numeric equations with functions and operators
Group Operations	Statistics across an analysis group such as min, max, average
Miscellaneous	Specialized types such as GPS Dilution of Precision (DoP)

4.5.5.1 Expression Analyses

The *Expression Analysis* combines other SOAP analyses (which can also be expressions) into either a logical or numeric result using operators and functions. Operators can be logical or numeric and include Boolean operations (AND, OR, NOT), conditions (IF, THEN, ELSE), relations (greater, less, equal), and arithmetic (addition, subtraction, multiplication, division, modulus, and exponentiation). Functions include common math library elements, such as trigonometric, logarithmic, and Bessel. Numerical computations have logical results based on the truth interval and are considered TRUE if the result lies between these two values.

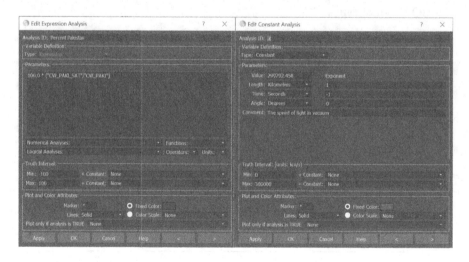

FIGURE 4.9   Expression analysis editor.

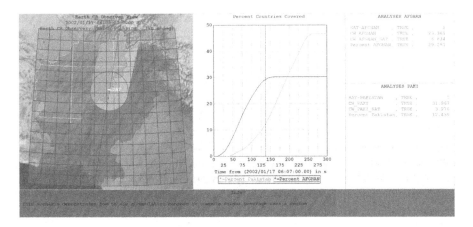

FIGURE 4.10   Expression analysis example with region sensors.

4.5.5.2   Constant Analyses

The *Constant Analysis* allows the user to define a numeric quantity having a specified set of distance, time, and angular units as exponents. As an example, the speed of light constant shown has a distance exponent of 1, and a time exponent of -1, resulting in a velocity. SOAP Expressions regulate the use of units to be conformant with normal mathematical expectations. For example, one cannot add quantities having differing units.

4.5.5.3   Example Analysis Output

The SOAP equation shown on the left side of Figure 4.9 is from the "region. orb" sample scenario file. The scenario and its output are shown in Figure 4.10. Two contours are set up, defining the region sensors comprising Afghanistan and Pakistan, respectively. A *Contour Weight* analysis is defined for each. These are SOAP miscellaneous analyses that return the percentage of pixels inside of each country, sampled from a surrounding SOAP rectangular *Contour Grid* (described later). The cumulative coverage of passing satellite sensor comprises another contour grid. The percentage of each country covered is the target of the SOAP expressions and is computed as 100 times the quotient of the satellite and country coverages.

The example also demonstrates the use of a SOAP palette and its associated views. The coverage is superimposed on a map using a world view in the upper left-hand port. An XY plot view is in the upper center port, with the dashed vertical line near the center denoting the current time. Instantaneous analysis results at the current time are shown on the right. At the bottom is a simple text view description of the scenario.

4.5.6   3D Models

3D Models provide the SOAP user with the means to place and orient CAD geometry of spacecraft, aircraft, ships, surface vehicles, and structures. The SOAP 3D model construct provides the user with the means to import *Surface-Based Model*

(SBM) files, bind them to SOAP coordinate systems, and attach them to SOAP platforms.

4.5.6.1   SBM Files

The SBM file is an ASCII format that has been created and optimized for SOAP. Constructing a complex model directly is generally not feasible. The preferred method for building models is to use a COTS CAD modeler to construct the model and then export it to the wavefront object/material (.obj/.mtl) format, which is widely used in 3D animation-based applications. The SOAP OBJSBM utility program then offers several options for transforming models to SBM format. The SBM file is composed of *Primitives*, each of which is addressable in terms of material properties and coordinate location and orientation. A *Node* is a platform-relative platform defining an inherited displacement from the model origin that is applied at pivot points for independent pointing and rotations.

4.5.6.2   3D Model Definition Panel

Interactively, 3D models are imported into a SOAP scenario using the 3D model definition panel shown in Figure 4.11. A block diagram shows how user inputs are mapped to the SOAP object architecture.

The SBM file input allows the user to select a file from the pull-down menu. Doing so will cause the model to be loaded into the scenario. SOAP uses a portable method for locating SBM files with files being stored in one or more of the following four locations:

- The invocation directory (the directory containing the active scenario file)
- The models' supplementary directory (as specified in the preferences menu directories panel)
- The SOAP directory (which contains the SOAP executable program)
- The SOAP "models" (the directory where all SBM files are located at SOAP installation)

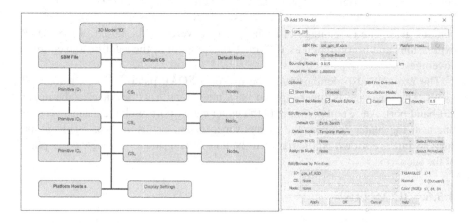

FIGURE 4.11   3D model definition.

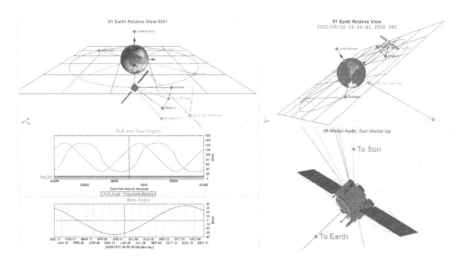

FIGURE 4.12   Visualizing spacecraft attitude.

SOAP searches for files in the order above. If separate model files having the same name are in multiple locations, the first model encountered is the one listed and loaded. This method and criteria are emblematic of how most external data files are managed by SOAP.

Mechanisms exist in the panel for assigning the 3D models to multiple platforms, and for assigning model primitives to SOAP coordinate systems. Usually, the entire file consists of coordinates normalized between -1 and +1. The entire model can then be scaled to a specified size using a single scale multiplier.

4.5.6.3   SOAP 3D Model Applications

One application of 3D models is visualizing spacecraft attitude. A scenario depicting the GPS yaw schedule is shown in Figure 4.12. In addition to seeing the model in various contexts, XY plots are used to characterize the yaw and beta angles. Yaw is a SOAP-included angle analysis, measured between the body X-axis and the orbit velocity vector. Beta is a SOAP beta angle analysis, measured between the orbit normal and the Sun position.

Figure 4.13 depicts a completely different application, though it also involves GPS. Shown is an SBM model of downtown Chicago.

FIGURE 4.13   Using 3D models for geospatial analysis.

In this application, the program wanted to assess constellation user receiver performance in urban canyons. The image on the left depicts the skyscrapers with real-time shadow computed. On the right is a contour-based elevation mask generated from a user position at intersecting streets. This mask is used by SOAP to occult the line on sight to the constellation, so the availability calculations would be constrained by real-world conditions. Once the mask is computed, it can be exported and used very efficiently without the CAD model being present. SOAP is presently being used in a similar fashion for contour UAS applications.

4.5.7 Trajectories and Swathes

Trajectories and swathes are both representations of past and/or future relative motion. A trajectory integrates platform position as a function of time relative to a reference platform and a coordinate system frame. A swath integrates planetary coverage of a sensor FOV across the surface of a planetary body. An image depicting trajectories and swathes is shown in Figure 4.14.

In the upper-left display trajectories shaped as figure-8s depict the relative motion between the primary satellite (I34) and those in the planes to the left (I24) and right (I44). Time markers are shown every minute of elapsed time. The graphs on the bottom are XY plots of SOAP analyses for relative position and velocity. In the right display, a sensor swath is shown with time markers every minute. The sensor has a conical scan pattern and is shown in the lower-left portion of the panel. The defining SOAP panels for the trajectory and swath examples are shown in Figure 4.15.

The trajectory object specifies a platform, reference platform, and coordinate system, and the swath a platform and a sensor. Both offer relative and absolute time interval and alignment settings, and step/marker settings. Both are generated objects, whose values are precomputed and stored. Recently, options for dynamic, instantaneous trajectories were added. The swath assumes a bounding cone about the sensor to prevent ambiguity.

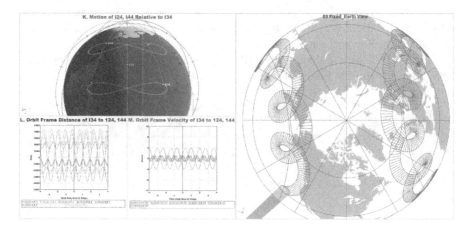

FIGURE 4.14 Trajectories and swathes.

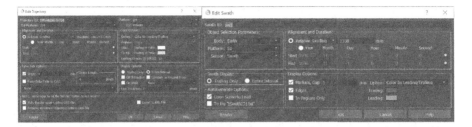

FIGURE 4.15 Defining panels for trajectories and swathes.

4.5.8 RF Transmitters and Receivers

Radio Frequency (RF) objects allow SOAP users to define communications properties of transmitting, receiving, and jamming terminals and investigate link closure under benign and interference scenarios. The models include the propagation effects of absorption due to rain and atmospheric gases. A functional block diagram is shown in Figure 4.16.

The receiver is the highest-level object and references primary and interfering Transmitters along with associated platforms, coordinate systems, and environmental parameters. Both transmitters and receivers have antennas as defining elements, with SOAP offering a library of International Telecommunication Union (ITU) recommendations as models. SOAP also supported one- and two-dimensional user-defined antenna gain patterns. A typical gain pattern for a parabolic dish is shown in Figure 4.17.

The SOAP RF module can produce numerous outputs, including link margin, received gain, received power, path loss, rain loss, and atmospheric loss. These computations are

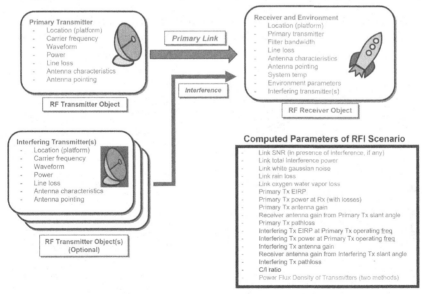

FIGURE 4.16 RF object architecture.

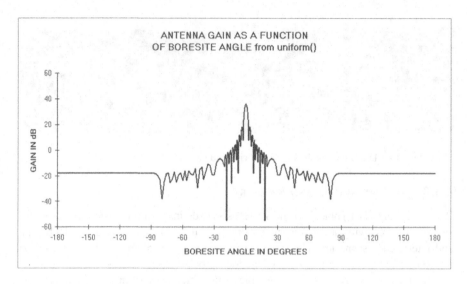

FIGURE 4.17 Gain pattern for a parabolic antenna.

represented as SOAP analyses. Some graphical outputs of the RF module are shown in Figure 4.18.

The left side of the palette shows extruded versus hemispherical contours of parabolic antenna gain. The large pane at the center shows a ground contour of the power flux density of a 2.2 GHz transmitter for a parabolic antenna.

4.5.9 ISOCHRONES

The *Isochrone* objects are used for signal processing applications. They represent lines of position based on Time or Frequency Difference of Arrival (TDOA or FDOA) of signals as received by two or more spacecraft. The inputs are a ground

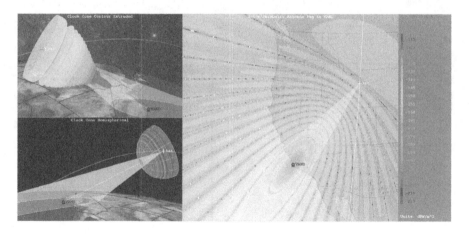

FIGURE 4.18 Graphical RF displays.

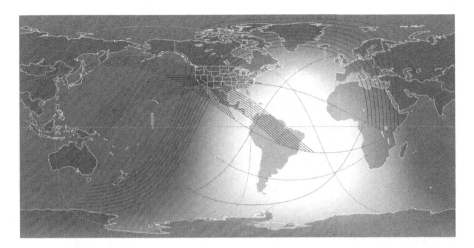

FIGURE 4.19   Time difference of arrival isochrones.

station platform suspected emitter (ground) location, and two or more satellite plat-forms. In the case of TDOA, multiple lines can be generated by the introduction of an ambiguity due to a pulse repetition frequency. In all cases, the curves are confined to the mutual coverage region of the participating satellites. An example of a TDOA plot is shown in Figure 4.19.

4.5.10   GROUPS

Groups are collections of objects that users can aggregate for various purposes. Included are platform, analysis, and terrain/imagery groups. Platform groups enable the user to edit platform display attributes in unison. They are also used as inputs and outputs for other SOAP object types and operations. Analysis groups allow the user to define variable platforms and to specify variable arrays for program input and output.

4.5.10.1   Platform Groups

Platforms Groups allow users to select an arbitrary set of platform for bulk editing of supported attributes, or for some analytical applications. Platform groups are auto-matically constructed by operations such as importing platforms from an external source (such as two-line element sets) file, and by defining satellite constellations using associated SOAP utilities. The set up for the general platform group is shown in Figure 4.20.

Member selection is shown on the right. Attributes that can be set across the membership are shown on the left and include toggling the display off/on, colors, icon types, and label rotation.

4.5.10.2   Platform Constellations

Platform constellations can be built with the utilities, *Build Constellation* option. Supported constellation types include *Walker, Streets of Coverage, Regular*, and *Flower*. An example of some flower constellations is shown in Figure 4.21.

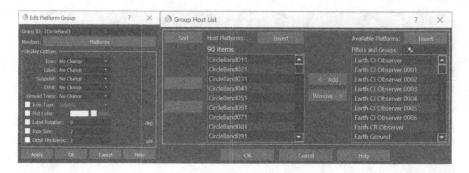

FIGURE 4.20   Platform group selection.

With all constellation types, the user specifies a total number of satellites. Other parameters include the number of planes, altitude, inclination, and phasing.

4.5.10.3   Analysis Groups

Analysis groups consist of a set or an array of individual SOAP analyses. This has been an area of active development targeting the analysis of emerging large satellite constellations. Analysis groups set the stage for automated definitions, and can also serve to reduce the number of objects in SOAP scenarios. Analysis groups can be constructed by selecting platform groups as arguments to SOAP analyses. They can also be constructed by selecting members from a list in an ad-hoc fashion. All members of an analysis group are expected to have the same SOAP unit type. A MATLAB Interface group is a special constuct designed to pass parameters between SOAP and extenal programs.

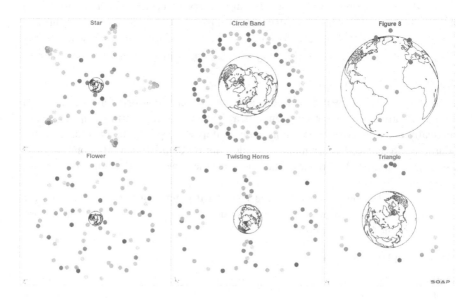

FIGURE 4.21   Six examples of flower constellations.

4.5.10.4   Group Operation Analyses

Group Operations Analyses operate across an analysis group and compute statistics such as group average, count, minimum, maximum, ordinal value, standard deviation, and sum. This sets the stage for the SOAP variable platform, which can assume the identity of platform group member having the Nth largest or smallest value of a specified computational metric.

4.5.10.5   Interface with the Variable Platform

The variable platform performs its selection based on an analysis group minimum, maximum, or an associated ordinal position, such as the fourth highest. The scenario depicted in Figure 4.22 depicts the satellite having minimum range to the satellite of interest, Iridium 98. The identity of the closest satellite and the range are disclosed and a link is drawn to it. The figure also includes an XY plot of range, and text values of the parameters of interest.

4.5.10.6   Terrain/Imagery Groups

Terrain/imagery groups allow the management of multiple *Terrain/Imagery Grids* (to be discussed in an upcoming section). These tile collections come from the SOAP imagery downloader utility program and other sources.

4.5.11   ACTIONS, CONDITIONS, DELTA-VS, AND THE SOAP SCRIPT

Actions are changes made to SOAP objects based on the occurrence of events resulting from the execution of a SOAP *Script. Conditions* are the criteria for applying Actions. The script is composed of a set of user-defined conditions and/or orbit transfers (delta-Vs). When the script is executed, it evaluates its constituent conditions

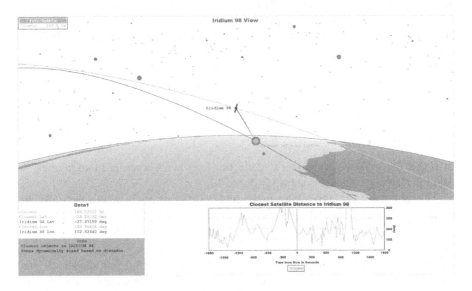

FIGURE 4.22   Analysis groups and the variable platform.

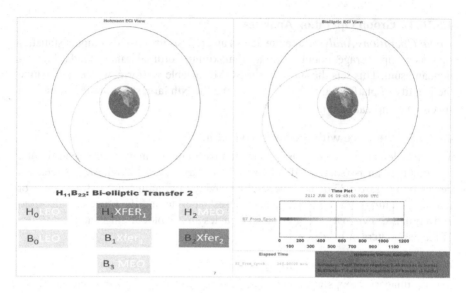

FIGURE 4.23   Script-based comparison of Hohmann and bielliptic transfers.

and delta-Vs. When the criteria are met, events are posted that carry out the actions at specific times. Together, the combined use of actions, conditions, events, and the script is called *Scripting*. The process of scripting requires the creation of all four of these constructs.

Figure 4.23 depicts the use of delta-Vs and the script to compare a Hohmann versus bielliptic transfer for the given orbit. A state diagram detailing the status of the orbits during the maneuvers is automatically updated as part of a SOAP *Sequence File*, which is embedded in a *Static Image View*. Also shown is the time window in an XY plot view, and the current time based on the color scale in an analysis data view. A text view displays the total amount of delta-V required for each orbit.

There is much more to SOAP scripting than can be presented in this brief introduction. The reader is referred to the SOAP user's manual and the example scenario files for more information.

4.5.12   DATA GRIDS

In SOAP, *Data Grid* objects refer to collections of values organized into meshes. A list of the supported types is shown in Table 4.7.

Although these grid types may apply to very different applications, the unifying principle is that they all deal with potentially large arrays of data arranged in grids.

4.5.12.1   Contour Grids

At their simplest, *Contour Data Grids* allow the user to define and render various labeled line and mesh shapes in SOAP world views. Such grids can be rendered on or above the planetary surfaces, or across planar or spherical sections oriented in 3D

TABLE 4.7
Data Grid Object Types

Data Grid Type	Description
Grid/Contour	Spatial lines, labels, with optional sampling of analysis values
Terrain/Imagery	Height fields based on external data products
Data Tables	Regular/irregular data value collections, up to four independent variables
External Program	Collections of parameter arrays for passing to MATLAB
Parametric Study	Variation of selected inputs in regularly sampled intervals

space. However, a dual function is the ability to sample numeric or logical SOAP analyses over these regularly sampled grids. As output, the contour grid displays color-coded values at each grid sample point. The contour is set up using the built-in ".Template Platform," a special wild card construct, which is moved to and evaluated at each grid sample point. SOAP supports numerous instantaneous and time-based statistical sampling modes and can compute grid-wide equal area weighted performance summaries.

The *Planet Contour Shape* is computed and rendered on or above the surface of a planetary body. Two such grids are shown in Figure 4.24. The grid on the left shows lines of latitude and longitude on a 3D projection of the Earth. On the right is the *Instantaneous Mode* coverage of a "Hosts of View Of" analysis from an associated satellite constellation, with the darker zones not being covered. At the bottom is a textual display of the percentage of Earth area covered.

Planet contour grids do not have to cover the entire surface; they can be limited between two longitudes and latitudes. Besides *Instantaneous Mode* SOAP offers a *Static Instantaneous Mode,* for data that do not change, such as terrain height. It also supports many different *Over Time* contour evaluation methods such as the *Age of Data Mode* shown in Figure 4.25.

The age of data mode computes how long it has been since each pixel has been visited by satellite coverage. These modes involve batch computation over a virtual time interval and can take minutes or even hours to generate. They can be saved to *Plot (.plt) Files*, which are then associated with the scenario and reloaded with it.

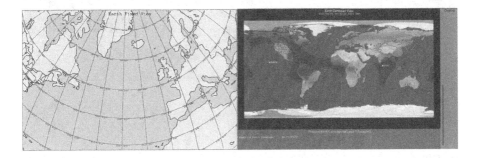

FIGURE 4.24 Planet grids with lines and visibility sampling.

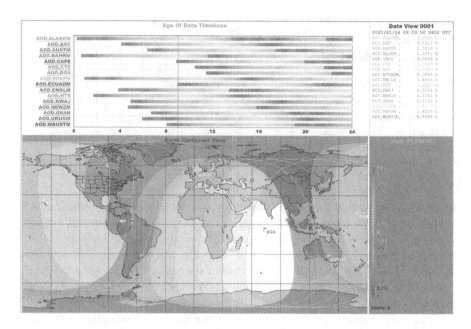

FIGURE 4.25   Age of data contouring mode.

A related contour shape is *Theta, Phi*. This is a coverage on all or a portion of a sphere. This can be centered on a planet, or another platform, such as a satellite. Theta is a measure in the XY plane of a coordinate frame, such as the Earth equatorial plane. Phi is measured above and/or below the plane. This harkens back to Figure 4.7 discussed in the context of Sensors. In Figure 4.26,

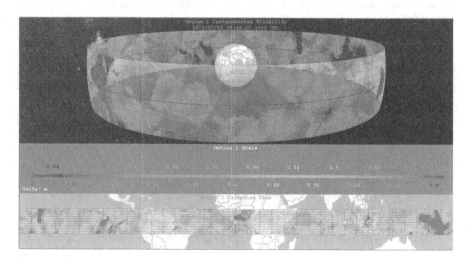

FIGURE 4.26   Theta-phi spherical contour grid.

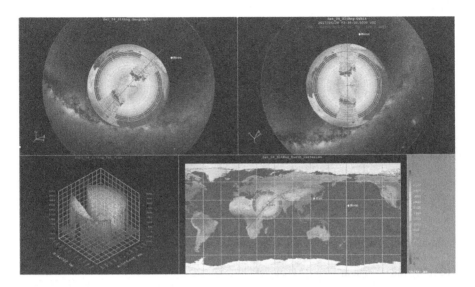

FIGURE 4.27   Radial cap contour grid.

a GPS accuracy measure such as DoP is plotted in the region above and below the satellite GEO belt.

The plots are shown in 3D and flat Earth (Cartesian) projections. Another useful grid shape is *Radial Cap* as shown in Figure 4.23.

This defines a circular or sectional area on or above a planet surface. The unique aspect of this contour type in SOAP is that it does not have to be planet-fixed and can roam the surface of the body as a satellite would. As such, it is an excellent way of showing payload coverage. The images in Figure 4.27 depict different views of a radial cap contour hosted on a LEO satellite and projected to the ground. Coverage is blocked by bus obscurations near the orbit plane, and the pixels there are masked off so that the scene background is visible. This plot is simply the slant range to the Earth surface, though more complex metrics can be set up. Views of the contour are shown in north-up, velocity-up, and Cartesian projections. A *Data Table Plot View* provides a 3D parameterized plot of the contour.

Contours can assume the shape of 3D Models. Figure 4.28 depicts a contour on the surface of a GPS spacecraft CAD model in simulated orbit.

3D model contours are sampled at vertex points. This example is low resolution, but the use of a SOAP tessellation utility can increase the sample density. A better contouring method would be to sample each surface polygon centroid and apply equal area weighting.

4.5.12.2   Terrain/Imagery Grids

The *Terrain and Imagery Data Grid* object type allows SOAP users to display terrain and/or detailed imagery on selected rectangular regions of planetary bodies. Terrain and imagery are often combined (fused). SOAP supports use tiled, monolithic, and

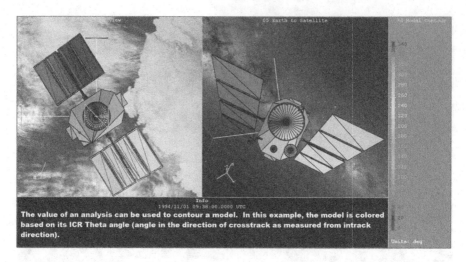

The value of an analysis can be used to contour a model. In this example, the model is colored based on its ICR Theta angle (angle in the direction of crosstrack as measured from intrack direction).

FIGURE 4.28   Contouring a CAD model.

CAD-based coverages. Tiled coverages are height field files covering $1 \times 1°$ extents with the locations encoded in the file names. These include

- *Digital Terrain Elevation Data (.DT0,. DT1,. DT2,. DT3)* from the National Geospatial-Intelligence Agency (NGA). 1 km, 100 m, 30 m, 10 m coverage, respectively
- *Shuttle Radar Topography Mission (HGT).* 1- and 3-arcsecond coverage
- *Grid Float Format* from *US Geological Survey (USGS).* SOAP does not support this directly, but employs a utility program, gf2hgt for converting individual grid float tiles to SRTM HGT format for subsequent loading and display. 10 m coverage

An example of Digital Terrain Elevation Data (DTED) coverage is shown in Figure 4.29. LANDSAT imagery has been superimposed over DTED Level 1. DTED 1 is classified as "LIMDIS" and is not released with SOAP. User agencies are expected to obtain their own geospatial data products.

Shown on the left of Figure 4.29 is a pair of F-18 SOAP 3D models over Southern California along with a hypothetical target on the ground. A static instantaneous contour elevation of terrain elevation is shown on the right, along with an associated legend in a *Color Scale View.* A plot of terrain elevation along the flight path is shown at the bottom.

SOAP can also import terrain data from individual files having arbitrary rectangular geographic extents. Usually, such coverages are used for higher resolution modeling of regions having smaller extents than for tiled terrain. These are packaged in data bearing GeoTIFF formats containing 16-bit integer or 32-bit floating point height data. Such data must be in WGS 84 (longitude, latitude) projections, as defined by the European Petroleum Survey Group (EPSG). The code for this projection is EPSG:4326. SOAP does not support the commonly used *Universal Transverse Mercator (UTM) Projection*, but Geographic Information System (GIS) packages such as GDAL or

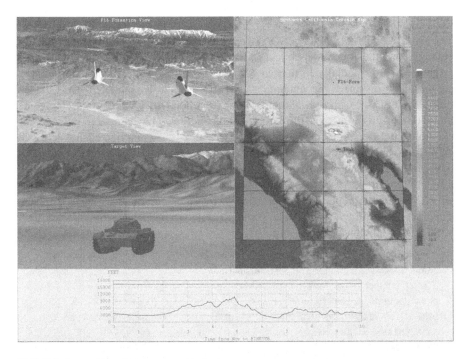

FIGURE 4.29   Terrain imagery grid display.

QGIS can be used to re-project (or "warp") UTM-based GeoTIFF coverages into EPSG:4326 Projections. When converting from UTM projections, it is recommended to merge all tiles into a single coverage before performing the reprojection.

SOAP *Data Table* files (ASCII or binary) can be used for terrain. To qualify, data tables must have two angular independent variables and a dependent variable that is in distance units. Terrain data can also be obtained in *Computer Aided Design (CAD)* formats. These formats are increasingly being used to publish products based on *Photogrammetry* collections. CAD-based models have some significant advantages over height-based models in that the data do not have to be regularly spaced or single-valued. With a CAD model, it is possible to represent objects with overhangs (such as bridges) and with vertical faces (such as buildings). SOAP uses 3D model objects rather than data grid objects to represent CAD-based coverages. Such models also can contain their own imagery collections. An example is the model of Chicago shown in Figure 4.13. The downside of CAD files is more data, as X, Y, Z coordinates must be stored for each vertex, and there can be overlapping vertices at different heights. Often, CAD files must be manually georeferenced.

A SOAP scenario can contain one or more potentially overlapping combinations of the terrain coverages as described above, When custom terrain files are present, SOAP will examine them for missing values and out-of-range indices. When such situations occur, SOAP will "fall back" to any underlying tiled coverage present in the scenario. This allows for a very high level of detail in the focus areas, and lower levels in the periphery.

4.5.12.3   Data Table Grids

The *Data Table Grid* enables SOAP analyses to index into *Data Tables* having between one and four independent variables and one dependent variable. These are loaded from specially formatted ASCII text or binary files from external data sources. For instance, the antenna gain patterns may be modeled using tables indexed by SOAP *Clock Angle* and *Cone Angle* analyses. Values that change as a function of mission elapsed time are good candidates for data table indices. For example, fuel depletion may be non-linear and/or too complicated for a SOAP expression but can be adequately approximated using a one-dimensional data table. In this case, the foreknown values of depletion are inserted into the table as a function of time. An *Elapsed Time Analysis* is then used to index into the table, which returns the depletion value. These are two examples of using data tables. Their scope is limited by the user's imagination. A simple one-dimensional data table is shown below.

```
TABLE_NAME FUEL DEPLETION EXAMPLE
DIM 1  COLUMN_NAME TIME
DATA_COLUMNS 6
DIST_UNITS KILOMETERS 0.0 TIME_UNITS SECONDS 1.0 ANGLE_UNITS DEGREES 0.0
DATA_UNITS kgs per min
DATA REGULAR
0      3600   7200   10800  14400  18000
12.2  10.1  7.5    5.5     12.1    9.8
```

This defines fuel depletion as a function of time shown in seconds. The header starts with the *TABLE_NAME* keyword, followed by an arbitrary name, "FUEL DEPLETION EXAMPLE."

Next is the *Dimension* keyword (DIM), which is followed by the integers -1, 1, 2, 3, or 4. This is the order (dimension) of the data to follow. For column-major format, this is set to 1, with -1 denoting row-major format.

The *COLUMN_NAME keyword* is the title of the data table's X axis. The name is for user reference only and does not affect user output. Likewise, the constructs ROW_NAME, LAYER_NAME, and STACK_NAME apply to the Y, Z, and T axes of 2-, 3-, and 4-dimensional tables. The column name here is TIME.

The *DATA_COLUMNS keyword* indicates the number of columns. For a one-dimensional table, this simply represents the number of data values. There are six indices in the table above. Tables with dimensions -1, 2, 3, and 4 include keywords for *DATA_ROWS*, *DATA_LAYERS*, and *DATA_STACKS*, respectively.

The DIST_UNITS, TIME_UNITS, and ANGLE_UNITS keywords define an exponent for each index. Only these unit types, or mixtures of them, are supported. As an example, velocity would be defined with a distance exponent of 1 and a time exponent of -1. Fractional values are supported. The *DATA_UNITS* keyword is a description of the units of the actual table data values (as opposed to the indices). The text provided here is for user reference only and does not affect the output. The example above indicates that the table output is in "kgs per min."

The *DATA* keyword is followed by REGULAR and IRREGULAR keywords. REGULAR is used if the indices are evenly spaced along each dimension.

Tables are bound to a SOAP scenario using data grids, data table selection of the SOAP Objects panel. The definition panel serves as a link between SOAP and the data table file. The interface allows users to browse, select, and load a data table file. Users can enter optional comments and an *Out of Range Value*. This value will be returned when *Index Analyses* exceed the range of the table. Indices are standard SOAP analyses having a matching unit with the corresponding data table dimension.

The *Data Table Analysis* provides the means for SOAP to extract values from data table grids (or contours or terrain grids). This is a *Miscellaneous* Analysis type in the SOAP objects panel. For data table grids, the inputs are a loaded data table object and one, two, three, or four analyses, depending on the dimensionality of the data. The inputs are analyses that serve as the data table indexes. These must be defined before the data table analysis can be completed. The output is a value from the table (or, optionally, an interpolated value from the table). This corresponds to the values of the input indices at the time the analyses are evaluated. The entire mapping of extraction from the fuel depletion data table is shown in Figure 4.30.

An important remaining topic is what happens when the index analyses of a data table analysis lie in between table indices. This is specified by the *Interpolation Method*. The *Closest Setting* searches for the nearest key index in the table and returns the data point corresponding to that index. In the example, a time index of 4,000 seconds would return 10.1. A time of 5,401 seconds would return 7.5. The *Linear Setting* enables *Linear, Bilinear, Trilinear,* or *Quadlinear* interpolation (depending on the dimension of the data), and the interpolated value is returned. This is shown in Figure 4.31 for a single dimension.

In the example shown in Figure 4.32, a time index of 4,000 seconds would return 9.8111 using linear interpolation. A time index of 5,401 seconds would return 8.7992. The *Retain Setting* holds the value of the dependent variable at the

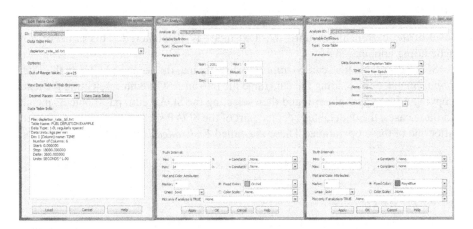

FIGURE 4.30 Returning values from a data table grid.

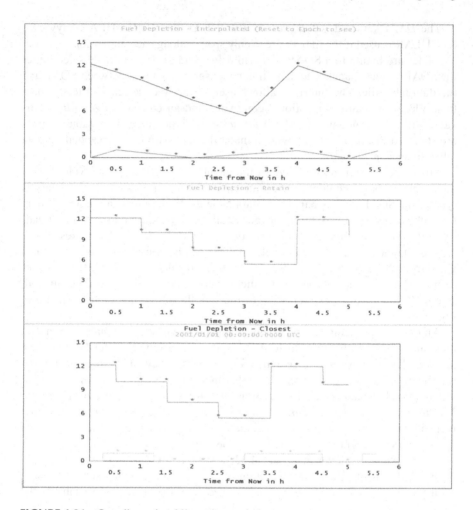

FIGURE 4.31 One-dimensional linear interpolation.

level of the current index until the next index is encountered. This is useful for scheduling applications.

SOAP data table files can be imported directly into the scenario by using the *File, Import* menu, or by loading the data table file into an ASCII editor, selecting the text, copying it into the clipboard, and then selecting the SOAP edit, paste objects menu. In these cases, the data tables become part of the SOAP scenario file once it is saved after one of these operations. These are called *Embedded Data Tables* and have the

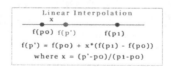

FIGURE 4.32 Data table interpolation methods.

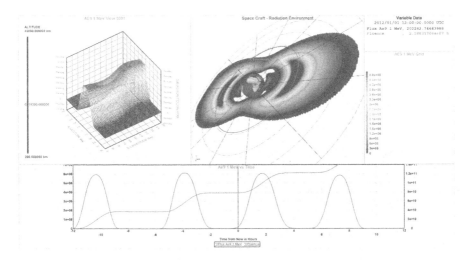

FIGURE 4.33   Data tables for trapped radiation particles.

advantage that they can be transported as part of the scenario. They operate the same way are external files, except that the file name field in the data grid, data table panel is set to <Embedded> and cannot be changed. The only way to unload such a data table is to delete the data grid object containing it.

Figure 4.33 visualizes a data table from the *Aerospace Electron and Proton Model* (AE9/AP9). This is a more sophisticate example with a data table having three independent variables (latitude, longitude, altitude). The dependent variable is electron or proton density. The visualization includes a data table plot view, with a slider for the off-screen dimension of altitude. To the right is a SOAP *Polar Planar Contour* with an orbit passing through it. The contour values change as the plane passes through different portions of the 3D grid. Instantaneous values and a color scale legend are shown on the right. The bottom display is an XY plot of instantaneous satellite exposure, or flux, and the integration of this for cumulative exposure, or fluence. Many different SOAP objects can conspire to present the big picture. The reader is referred to the SOAP user's manual for detailed descriptions of higher dimension data tables.

4.5.12.4   External Program Grids
External Program Grids allow users to gather sets of parameters for passing to and from external software such as MATLAB. SOAP analyses groups are set as inputs or outputs. An analysis group must be used for each parameter, even when it is a scalar.

Analysis groups are selected, and the user can re-order the list. The SOAP analysis groups used to house the output usually contain an array of *Interface Analyses*. These are generic placeholders that can accept data from an external source.

4.5.12.5   Parametric Study Grids
The parametric study data grid object allows the user to generate sets of varied inputs for constructing parametric studies. An example is shown in Figure 4.34.

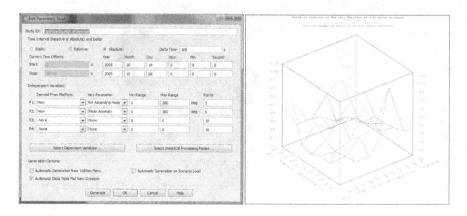

FIGURE 4.34 Parametric study data grid.

The results of these are generated in a SOAP data table file. After generation is complete, SOAP automatically loads the data table file into a corresponding SOAP data table grid object and displays it in a data grid plot view. In this case, two independent variables are varied, and a two-dimensional data table is produced. The text is shown in Table 4.8.

TABLE 4.8
Data Table Produced by a SOAP Parametric Study

```
TABLE_NAME Contour weight of US PDOP Average CW
COMMENT "Study duration from 2005 OCT 19 00:00:00.00 UTC to 2005
OCT 20 00:00:00.00 UTC"
DIM 2
COLUMN_NAME RA Ascending Node
COLUMN_COMMENT "RA Ascending Node of New varies from 0.000000 to
300.000000"
DATA_COLUMNS 6
DIST_UNITS KILOMETERS 0
TIME_UNITS SECONDS 0
ANGLE_UNITS DEGREES 1
ROW_NAME Mean Anomaly
ROW_COMMENT "Mean Anomaly of New varies from 0.000000 to 360.000000"
DATA_ROWS 7
DIST_UNITS KILOMETERS 0
TIME_UNITS SECONDS 0
ANGLE_UNITS DEGREES 1
DATA_UNITS None
DATA REGULAR
```

	0.000000	60.000000	120.000000	180.000000	240.000000	300.000000
0.000000	0.640546	0.295635	0.302116	0.544636	0.655098	0.391922
60.00000	0.276412	0.292272	0.478170	0.521394	0.900993	0.329033
120.0000	0.294007	0.278413	0.351637	0.638955	0.314413	0.281777
180.0000	0.247979	0.329520	0.508876	0.448489	0.329579	0.260632
240.0000	0.282478	0.460487	0.321565	0.400152	0.266290	0.264440
300.0000	0.779572	0.434241	0.404824	0.272240	0.444015	0.506095
360.0000	0.640546	0.295635	0.302116	0.544636	0.655098	0.391922

FIGURE 4.35 SOAP menu system.

4.6 OTHER MENU OPTIONS

The SOAP menu system as shown in Figure 4.35 is accessed through the menu bar.

There are many topics to discuss, particularly in the utilities and preferences menus. Detailed descriptions and examples of the entire set of SOAP features come with the software.

4.7 SUMMARY

SOAP is a general-purpose orbital and geospatial modeling tool developed at the Aerospace Corporation. It offers an open, flexible architecture that is predicated on user creativity and innovation. The software is available for work on U.S. government contracts. Training is available, with hands-on tutorials and a detailed user manual included in the program distribution.

ACKNOWLEDGMENT

The Aerospace Corporation. Approved for Public Release OTR 2023-00181

REFERENCES

D. Y. Stodden, & G. D. Galasso, *Space System Visualization and Analysis Using the Satellite Orbit Analysis Program (SOAP)*, IEEE Aerospace Applications Conference Proceedings Vol 2, Aspen CO, 1995.

D. Y. Stodden, J. Coggi, J. A. Paget, M. P. Phan, & R. L. Swartz, *SOAP 15.3.x User's Manual*, Aerospace Corporation OTR#20130318072637, 2021.

J. Coggi, & D. Y. Stodden, *Introduction to the Satellite Orbit Analysis Program (SOAP): Part 1 in the SOAP Course Framework*, Aerospace Corporation OTR#20090204133333, 2022.

J. Coggi, & D. Y. Stodden, *What is Aerospace's Satellite Orbit Analysis Program (SOAP)?*, YouTube video, https://www.youtube.com/watch?v=nq_q5Qh2lBE [November 12, 2020].

5 Advance Framework for Simulation, Integration, and Modeling (AFSIM) and Its Space Capabilities

Larry B. Rainey

NOTE TO READER

This chapter covers new and emerging technologies in the field of space warfare analysis. At the time of publication, there is limited information or accessible literature on this topic. As such, the chapter has been compiled by the editor from publicly available articles, fact sheets, and other online sources [1].

All sources in this chapter are in the public domain or reprinted with permission.

SECTION 1: OVERALL AFSIM GENERAL DESCRIPTION

Note, the following text comes from the article entitled "AFSIM: The Air Force Research Laboratory's Approach to Making M&S Ubiquitous in the Weapon System Concept Development Process" [2].

* * * *

5.1 INTRODUCTION

Arguably, one indication that a burgeoning technical concept or capability is on the precipice of widespread usage and acceptance is when it enters the United States (U.S.) Congressional record. For Digital Twin, the notion of building realistic digital models of real-world systems, that moment has arrived. The House Armed Services Committee's Subcommittee on Tactical Air and Land Forces recently drafted language for the Fiscal Year 2020 National Defense Authorization Bill directing the U.S. Secretary of Defense to provide a briefing to the committee explaining "how the F-35 program is implementing the use of digital twinning technology across the F-35 system enterprise" [3].

In order to mainstream the effective implementation of "digital twinning" of F-35 and other weapon systems, the U.S. Department of Defense (DoD) must have a modeling framework that is effective, available, affordable, and relatively easy to use. DoD must also have a culture that is accepting of the results produced by these digital models. The Air Force Research Laboratory (AFRL) is making clear headway on both fronts with its Advanced Framework for Simulation, Integration and Modeling (AFSIM), a C++-based, modular, object-oriented, multi-domain, multi-resolution

DOI: 10.1201/9781003321811-7

modeling and simulation (M&S) framework for military simulations focused on analysis, experimentation, and wargaming. The AFSIM community already encompasses over 1,200 trained users across 275 organizations, including all branches of the U.S. military; other U.S. government agencies; industry; academia; and our five closest allies. This broad-based community is already widely using AFSIM to assess and compare various weapon system concepts, refine operational employment tactics for the most promising concepts, and ultimately to inform the weapon system investment decisions within AFRL and across the DoD. This paper describes the steps AFRL is taking and the progress achieved in making AFSIM as ubiquitous in the defense M&S community as MATLAB is in the academic community.

5.2 AFSIM 101

In its present form, AFSIM represents a government/industry investment in excess of $50M. Between 2003 and 2013, Boeing invested approximately $35M of Independent Research & Development (IR&D) funding into what it called the Analytic Framework for Network-Enabled Systems (AFNES) that Boeing designed to simulate threat-integrated air defense systems (IADS). Frustrated with the proprietary, inflexible M&S tools available to the government at the time, AFRL conducted a head-to-head showdown of available tools in 2011, selecting AFNES as the framework of choice for its trade-space analysis and technology maturation M&S work. In 2013, Boeing transferred AFNES to AFRL with unlimited rights, which AFRL subsequently rebranded as AFSIM [4].

Note that the "AF" in the AFSIM name does not stand for Air Force. This reflects AFRL's belief that AFSIM should not be just an internal Air Force tool, but rather a common framework used broadly across the entire defense M&S community. This naming choice also signifies that AFSIM is more than just a framework for simulating aircraft. It was designed to be a multi-domain platform, meaning it can model land, sea, air, and space-based platforms, enabling modelers to include submarines, naval vessels, tanks, airplanes, helicopters, satellites, and even cyber agents in the same simulation, if needed.

From its earliest conception, AFSIM was also envisioned to be an open system, utilizing "plug and play" modules to overcome the expansion and compatibility constraints of earlier frameworks. This modular approach allows the modeler, rather than the AFSIM programmer, to determine the appropriate level of fidelity (i.e., the degree to which the underlying physics are simulated) for the models used in the simulation. Likewise, users can adjust the fidelity of each platform to meet their specific simulation needs. The fidelity of an airplane model, for instance, could vary between a point in space moving along a predefined vector to a full six degrees of freedom model that changes speed, direction, altitude, etc., based on the displacement of the virtual cockpit controls. The modular approach also enables the reuse and/or modification of existing models of various platforms without changing the core AFSIM code.

AFSIM's modular structure enables AFRL to distribute the code at two security classification levels, which users can adapt to meet their specific security requirements by adding additional software modules. AFRL offers both an unclassified and classified (U.S. Secret) variant of the code. The primary difference between the two available variants is simply the number, type, and fidelity of included models. To receive the classified variant of the software, contractors must also provide a current, certified DD

Form 254 Contract Security Classification Specification. The classified version also comes standard with National Air and Space Intelligence Center (NASIC) approved models of many threat systems, and a National Geospatial-Intelligence Agency (NGA) Digital Terrain Elevation Data (DTED) model. End users may then add their own modules to incorporate models of other platforms of interest for their specialized use. The overall classification of a given instantiation of AFSIM is then driven not only by the initial variant of the software, but also by the classification of modules added to that instantiation. Because of their inherent military utility, both variants are subject to International Traffic in Arms Regulations (ITAR) restrictions, meaning individuals and organizations can be fined or prosecuted for unauthorized release or export of the software. Because of these restrictions, academic institutions must have an approved, ITAR-compliant environment before AFRL can release AFSIM to them. Despite this restriction, the pool of academic users is growing. Georgia Tech, Purdue, Ohio State, University of Central Florida, University of Alabama in Huntsville, and the University of Illinois Urbana-Champaign are already part of the AFSIM family.

AFSIM spans a broad spectrum of military simulations to include the engineering, engagement, mission, and "campaign-lite" level via analytic wargaming and experimentation. As Table 5.1 depicts, the Engineering level consists of short-duration subsystem interaction with other subsystems. One example of this could be a radio frequency (RF) transmitter interacting with a receiver to identify subsystem-level capabilities and limitations. The Engagement level consists of "mano a mano" combat, that is, a brief exchange between two entities, or platforms, in the AFSIM vernacular. For instance, a missile exchange between a Blue (Friend) and Red (Foe) aircraft would constitute an engagement-level simulation. The next level of complexity would be the Mission level, simulating, for example, a series of combat exchanges between multiple Red and Blue aircraft over the duration of a single sortie or mission, nominally a few hours. These simulations can contain up to thousands of entities. Campaign-level engagements extend this even further, potentially including all the Red and Blue platforms in a given area over an extended period (i.e., days or even months). The focus for AFSIM development has been primarily at the engagement and mission level, with recent development expanding AFSIM to include "campaign-lite" capabilities via analytic wargaming. Other M&S tools are leveraged when needing to fully explore engineering or full campaign modeling, such as the Synthetic Theater Operations Research Model (STORM) used by the Air Force Studies, Analyses and Assessments Office (AF/A9).

AFSIM enables its user to scale the scenario to the appropriate simulation level to best study the item(s) of interest. Each subsequent level logically builds on the lower

TABLE 5.1
Levels of Wargaming Simulations

Simulation Level	Complexity Scale	Time Scale
Campaign	Many vs. Many	Days
Mission	Several vs. Several	Hours
Engagement	One vs. One	Minutes
Engineering	Subsystem Interaction	Seconds

levels to create a more intricate simulation in order to identify system-of-systems emergent properties that may not be apparent in simpler simulations. For instance, the combat effects of depleting munitions and fuel reserves may be unnoticeable at the Engagement or Mission level, but a total game-changer in the Campaign-level simulation. The limiting factor in the size and complexity of an AFSIM simulation is the storage, memory, and computing power of the host platform—and the associated wall clock time required to run the simulation. AFSIM allows users to set the desired balance between processing time and output fidelity by adjusting the various parameters and behaviors associated with the platform.

To achieve the degree of flexibility in platform type, fidelity, and simulation type described above, AFSIM uses four architectural elements (attributes, information, components, and links) to describe each platform in the simulation, as depicted by Figure 5.1. Attributes include standard data as platform name, type, and affiliation. This sub-element can be expanded to include mission-unique information such as radar, optical, and infrared signature data to determine an aircraft's vulnerability to detection by enemy sensors. The information element encompasses data resident on the platform, along with details on how these data are perceived by the humans that receive them. For an aircraft, this would include the sort of data that would be displayed to the pilot (i.e., altitude, speed, heading, radar indications, etc.), along with the myriad raw data driving these displays. The components element consists of various models that directly control how the platform behaves. These models describe how the platform moves through space-time, senses the surrounding environment, processes the information it collects, communicates with other platforms, and employs its arsenal of kinetic and non-kinetic weapons against adversary platforms, and conducts various other tasks. Finally, the links element coordinates the data exchanges between various subsystems on the platform, as well as communications with other platforms.

Another notable aspect of AFSIM is its support of both virtual and constructive simulations. In a constructive simulation, simulated operators control simulated systems—such

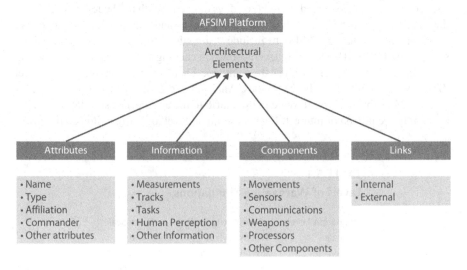

FIGURE 5.1   AFSIM architectural elements.

as a military battle where the Red and Blue players are all computer-controlled. In a virtual simulation, you have real operators controlling simulated systems—such as a pilot flying a flight simulator. AFSIM can be used constructively to conduct large trade space exploration of military capabilities, potentially involving tens of thousands of unique test points executing in a non-real–time manner. The results of such constructive simulation activities can then be utilized to define and conduct a virtual simulation that runs in real time to investigate a narrower trade space (informed by the constructive simulation) for a more focused assessment with operational pilot participation. This allows the same underlying simulation models to be utilized in both the constructive simulation and in the virtual simulation, providing more consistent modeling and analysis across both environments. In addition, AFSIM can also be linked into other simulations or other simulators/emulators to provide a true Live-Virtual-Constructive (LVC) simulation capability. Using Distributed Interactive Simulation (DIS) or other supported communication protocols, AFSIM can interact with other simulations or live experiments in order to provide additional entities (both virtual and constructive), system and subsystem models, threat systems, or potentially other simulated capabilities. This allows AFSIM to augment and/or complement a larger simulation or experimentation environment with additional capabilities as needed to best achieve any given test and analysis objectives.

Leveraging its Warlock graphical user interface (GUI), AFSIM likewise allows "operator in the loop" execution to facilitate analytic wargaming. Specifically, Warlock> enables operators to trigger various scenario events, control individual platforms, and even experience the mission from inside the platform—much like Flight Simulator. Warlock also facilitates the creation of "cells" of operators, e.g., a Blue Cell of friendly platform operators and a Red Cell of adversary platform operators, each of which only has access to virtual information collected by that cell's platform. In other words, the Red Cell and the Blue Cell each has imperfect information about the other. AFSIM can add further realism to the wargame by degrading the flow of information between members of the same cell. Warlock also supports the creation of a White Cell—the referees—that have perfect information about all platforms, which they leverage to control the flow of the overall wargame.

Before any type of AFSIM simulation can be executed, however, the user must define the various platform and component models and then craft the wargame scenario. To facilitate this process, AFRL distributes Wizard, AFSIM's Integrated Development Environment (IDE), as a supporting tool (like Warlock). Much like a modern software development IDE, Wizard serves as a single application to edit scenario files; write AFSIM script; graphically manipulate scenario laydowns; run software executable (i.e., run the scenario in AFSIM); and view the resulting output or error messages. It also highlights file syntax, flags unknown commands, and provides context-sensitive documentation. Wizard even comes with an auto-completion feature and a script debugger to minimize the time required to develop and debug models and scenarios.

5.3 THE ROAD TO UBIQUITY

Recognizing that widespread adoption is more probable by leveraging incentives rather than with mandates, AFRL has taken several steps to grow the AFSIM following. First, AFRL decided it would give away the product to both government and industry partners. Intra-government sharing could easily be accomplished under Memoranda of Understanding (MoU). However, sharing software with industry partners initially proved tricky, as existing contract mechanisms only allowed the sharing of government property and information with industry partners as part of a larger contract. The F-22 aircraft program could loan Lockheed-Martin the software as part of the larger F-22 contract, but the rule-set associated with government-furnished property meant the software could only be used for M&S work within the scope of the F-22 contract, and that the software must be returned to the government at the conclusion of that contract. To overcome this obstacle, AFRL created a new type of contractual agreement, an Information Transfer Agreement (ITA), that gives industry partners full access to the software, without constraining its use to a single program [5, 6].

AFRL also chose to provide free training at its Dayton, Ohio, headquarters. AFRL currently offers two courses: one for general users and one for code developers. The user course is offered monthly while the developer course is offered every other month. The only costs to attendees are travel-related expenses. This combination of free software and free training makes AFSIM very attractive to organizations that might otherwise be forced to use costly commercial off-the-shelf (COTS) products, along with the recurring expenses associated with license renewals, specialized training, and product support.

To further sweeten the deal, AFRL also opted to provide users and developers with the source code for both the framework and all supporting tools. This decision was borne from the lab's own frustration with other "black box" software tools that provided limited insight into how the tool transformed inputs into outputs. AFRL recognized that providing source code would enable savvy users to see for themselves the logic, algorithms, equations, and associated assumptions behind every AFSIM result. AFRL also realized that source code access could leverage the user community as code debuggers, knowing that inquisitive users would likely dig into the source code to understand anomalous results, unearthing logic errors and faulty assumptions that could be corrected in future software updates.

Deliberate community engagement has also been a core element of AFRL's strategy for AFSIM. In addition to actively soliciting feedback on the user experience and leveraging them to find and repair minor coding issues, AFRL has incorporated the user community into its governance model, establishing eight domain-centric working groups (Sensors, Space, Threats & Scenarios, Kinetic Weapons, Directed Energy, Standardization, Virtual Simulation and Wargaming, Cyber/C3) to help establish the vision for capability development within each group's respective domain. Each working group is a self-organized entity whose leadership structure is driven more by consensus of the subject matter experts in that group rather than by AFRL dictate. Nearly half the groups are led by non-AFRL personnel, some by other military services. A central program management team integrates and prioritizes the inputs from each group in order to develop an annual execution plan within the available funding limits.

As the AFSIM community has grown, so, too, has the need to scale the AFSIM software development and maintenance effort. The AFSIM software team now consists of over 40 full-time developers and analysts to maintain both the Windows and Linux variants. Software increments for both variants are released on a six-month cycle, with user support and bug fixes provided for both the latest version and one prior version. This approach enables a steady flow of cutting-edge capabilities, while providing longer-term stability for users who do not require the latest release. Each release of AFSIM has a one-year support window. As of this writing, AFSIM 2.3 is the "stable" version, AFSIM 2.4 is the latest version, and 2.5 is in development. The AFSIM development team works closely with network approval authorities to ensure authority to operate on multiple government systems across a range of security classifications. To facilitate these network approvals, the development team utilizes a continuous integration and build process that incorporates automatic builds, static code analysis, regression testing, and rigorous vulnerability scanning.

One of AFSIM's primary use cases is as a simulation platform for technology maturation. Over the past few years AFRL has made a significant investment in using AFSIM as a test bed for maturing air vehicle autonomy. Utilizing AFSIM as a simulation test bed for autonomy has created a single, unified environment for developing, maturing, and testing autonomy algorithms for basic and applied research, as well as for advanced applications. Using AFSIM as a virtual test bed for accelerating air vehicle autonomy development has proven so effective that several government agencies and industry partners have also adopted it for similar efforts, including the Defense Advanced Research Projects Agency (DARPA), Johns Hopkins University Applied Physics Laboratory (JHU APL), Georgia Tech Research Institute (GTRI), and Leidos. AFRL is also teaming with the Air Force Lifecycle Management Center (AFLCMC) and the Air Force Warfighting Integration Center (AFWIC) to make AFSIM the tool of choice for analyses of alternatives (AoAs) for future weapon system concepts. Additionally, AFWIC has incorporated AFSIM into its capability development guide. AFRL has also communicated to its industry partners that AFSIM will be a key tool used to evaluate their proposals. Lockheed-Martin's recent announcement that it is investing $5M into their AFSIM infrastructure is a clear indicator that industry is listening. Boeing, who developed the predecessor to AFSIM, has been a committed user for over a decade.

CONCLUSION

Representing a $50M investment to date and another $6M per year for the foreseeable future, AFSIM is not an inexpensive framework. However, AFRL believes the DoD will ultimately recoup this investment by reducing schedule delays and the associated cost overruns through earlier identification and correction of murky requirements, invalid assumptions, and flawed design decisions. Despite its shortcomings, AFSIM is already enhancing the DoD's "model centric" approach to acquisition. AFRL's conscientious efforts to make AFSIM useful, available, affordable, and user friendly have undoubtedly helped in this regard. AFRL believes that AFSIM will be key to helping the Secretary of the Air Force attain her vision of building an innovative Air Force that "dominates time, space, and complexity in future conflict across all operating domains to project power and defend the homeland" [7]. Will AFSIM ultimately help the Air Force achieve this lofty goal? Only time will tell.

SECTION 2: SPACE SYSTEMS ASPECTS OF AFSIM

As was stated in Section 1, "AFSIM is an engagement and mission-level simulation environment written in C++ originally developed by Boeing and now managed by the Air Force Research Laboratory (AFRL). AFSIM was developed to address analysis capability shortcomings in existing legacy simulation environments as well as to provide an environment built with more modern programming paradigms in mind. AFSIM can simulate missions from subsurface to *space* (emphasis added) and across multiple levels of model fidelity."

This section drills down to specifically address the specific space systems aspects of AFSIM as found at YouTube citation [8].

* * * *

CONCLUSION

This chapter addresses the tool called AFSIM and its associated space capabilities. It is both an engagement and mission-level simulation that has been written in the C++ language. It was developed to address shortfalls in legacy simulation environments with associated up-to-date programming methodologies. Its scope reaches from subservice all the way to space in context of multiple levels of fidelity. The AFSIM environment has three software pieces (i.e., the framework itself, an integrated development environment, and a visualization tool). An agent modeling architecture is also used to address behavior tree and hierarchical tasking.

ACRONYMS

AF/A9	Air Force Studies, Analyses and Assessments Office
AFLCMC	Air Force Lifecycle Management Center
AFNES	Analytic Framework for Network-Enabled Systems
AFRL	Air Force Research Laboratory
AFSIM	Advanced Framework for Simulation, Integration and Modeling
AFWIC	Air Force Warfighting Integration Center
AoA	Analysis of Alternative
COTS	Commercial Off The Shelf
DARPA	Defense Advanced Research Projects Agency
DoD	Department of Defense
DIS	Distributed Interactive Simulation
DTED	Digital Terrain Elevation Data
GTRI	Georgia Tech Research Institute
GUI	Graphical User Interface
IADS	Integrated Air Defense Systems
IDE	Integrated Development Environment
IR&D	Independent Research & Development
ITA	Information Transfer Agreement
ITAR	International Traffic in Arms Regulations
JHU APL	Johns Hopkins University Applied Physics Laboratory
LVC	Live-Virtual-Constructive
M&S	Modeling and Simulation
MoU	Memoranda of Understanding
NASIC	National Air and Space Intelligence Center
NGA	National Geospatial-Intelligence Agency
RF	Radio Frequency
STORM	Synthetic Theater Operations Research Model
US	United States

REFERENCES

1. AFSIM Abstract. See worldcomp-proceedings.com/proc/p2015/CSC7058.pdf [Accessed: September 19, 2023].
2. AFSIM Article Itself. See https://csiac.org/articles/afsim-the-air-force-research-laboratorys-approach-to-making-ms-ubiquitous-in-the-weapon-system-concept-development-process/ [Accessed: September 19, 2023]. The following five references come from the AFSIM article in referenced here.
3. U.S. House of Representatives, "H.R. 2500 – FY20 National Defense Authorization Bill – Subcommittee on Tactical Air and Land Forces," Washington DC, June 2019.
4. P. D. Clive et al., "Advanced Framework for Simulation, Integration and Modeling (AFSIM)," p. 5, 2015.
5. L. Daigle, "AFRL Uses New Information Transfer Agreement to share software with industry," Military Embedded Systems, 2017. Available: http://mil-embedded.com/news/afrl-uses-new-information-transfer-agreement-to-share-software-with-industry/ [Accessed: July 7, 2019].
6. J. Knapp, "Information Transfer Agreement Enables AFRL Software Sharing with Industry," Wright-Patterson AFB, March 10, 2017. Available: http://www.wpafb.af.mil/News/Article-Display/Article/1109831/information-transfer-agreement-enables-afrl-software-sharing-with-industry [Accessed: July 7, 2019].
7. US Air Force, "U.S. Air Force Science and Technology Strategy," April 17, 2019.
8. Team AFAMS, "Advanced Framework for Simulation, Integration and Modeling (AFSIM)," YouTube, 2019. See https://www.youtube.com/watch?v=8DMY5leZ2kA&t=5s

6 Extended Air Defense Simulation (EADSIM)

Larry B. Rainey

NOTE TO READER

This chapter covers new and emerging technologies in the field of space warfare analysis. At the time of publication, there is limited information or accessible literature on this topic. As such, the chapter has been compiled by the editor from publicly available articles, fact sheets, and other online sources.

All sources in this chapter are in the public domain or reprinted with permission.

6.1 BACKGROUND

This chapter contains a short article on EADSIM [1] and a publicly provided fact sheet developed by the United States Army Space and Missile Defense Center Public Affairs Office in Huntsville, AL to quote from. It is classified as Distribution A 0223-06 [2].

As stated in Chapter 1, I personally have used such tools as Systems Tool Kit (STK), Satellite Orbit Analysis Program (SOAP), System Effectiveness Analysis Simulation (SEAS), Extended Air Defense Simulation (EADSIM), and Synthetic Theater Operations Research Model (STORM) for the sake of conducting space system mission analysis. Using such tools as the first three provided me the requisite information needed to be employed in such engagement/mission and campaign tools such as EADSIM and STORM models respectively to see the impact of space-derived information at both the engagement/mission and campaign levels of conflict.

* * * *

6.2 U.S. ARMY FACT SHEET ON EXTENDED AIR DEFENSE SIMULATION (EADSIM)

Extended Air Defense Simulation, or EADSIM, is a system-level simulation of air, space, and missile warfare developed by the U.S. Army Space and Missile Defense Command Space and Missile Defense Center of Excellence's Capability Development Integration Directorate. It is used for scenarios ranging from few-on-few to many-on-many, and it represents a broad range of missions on both sides.

Each platform (such as a ground-based missile defense interceptor) is individually modeled, as is the interaction among platforms, their sensors, their launchers, and battle managers. EADSIM models the command and control decision processes and

DOI: 10.1201/9781003321811-8

the communications among platforms on a message-by-message basis. Intelligence, surveillance, and reconnaissance is explicitly modeled to support offensive and defensive applications.

EADSIM models fixed and rotary-wing aircraft, tactical ballistic missiles, cruise missiles, infrared and radar sensors, satellites, command and control structures, sensor and communications jammers, communications networks and devices, and fire support in a dynamic environment that includes the effects of terrain and attrition on the outcome of the battle.

Object behavior is controlled via flexible rulesets. This is the primary means for modeling battle management in EADSIM. Users select rulesets, behaviors, set parameters in the rule-sets, and program trigger event/response combinations to control the dynamic reactions of platforms to events in a scenario. Multiple rule-sets are available for each of the following categories: airbases, aircraft, defensive commanders, offensive commanders, sensor platforms, and surface platforms. Hierarchical, distributed, and cooperative relationships are modeled. Many facets of attack operations, including intelligence, surveillance and reconnaissance, target engagement, and battle damage assessment are modeled in EADSIM.

EADSIM provides a number of distributed simulation, operational planning, exercise, training and wargaming interfaces. These include distributed interactive simulation, high level architecture capability, and tactical communications. Other interfaces include Playback Interactive Console, an independent, graphics-based program used to review scenario results in an animated battle scene format; Force-on-Force Interactive Retasking Environment for operator-in-the-loop interactions supporting exercises; operational planning tools for defense analysis; and a number of operational database interfaces for increased fidelity.

EADSIM is used extensively for analyses of alternatives, integrated air and missile defense and intelligence, and surveillance studies.

- Nearly 300 user communities worldwide
- Supports commanders, combat developers, trainers, testers, and analysts in a single package
- Significant contributor to homeland defense/homeland security in point defense studies
- Supports more than 30,000 individual players in faster than real time
- Supported with ongoing maintenance, documentation, user group meetings and telephone hotline
- Active defense
  - Surface-to-air engagements
  - Air-to-air engagements
  - Multi-tier engagements
  - Theater ballistic missile engagements (boost, midcourse, terminal phases)
  - Surface-to-surface engagements (cruise, ballistic, advanced)
- Passive defense
  - Infrared signature
  - Radar signature

- Attack operations
  - Surface-to-surface attacks
  - Air-to-surface attacks
  - Surveillance
  - Intelligence collection
- BM/C3I
  - Engagement logic
  - Command and control structure
  - Communications networks
  - Protocols
- Integrated air and missile defense
  - Preferential selection of sensors to provide launch on remote/engage on remote
  - Centralized control/distributed engagement coordination/decisions
  - Configurable information exchange between sensors, command and control nodes, and shooters
  - Imperfect correlation/miscorrelations/lack of correlation/over-engagement/under-engagement of tracks
- Cyber electromagnetic activities
  - Downstream effects on combat systems
  - Cyber-attack timing fixed, randomized, or tied to dynamic simulation event
  - Agile jammers to prioritized targets
  - Multiple jammers stacked on threat emitter [1].

* * * * *

6.3 CONCLUSION

EADSIM is the most mature and widely used force-on force model in the world. With almost 30 years of development and high fidelity refinements, EADSIM is the preferred model for evaluation and validation of a broad range of weapon systems and capabilities. EADSIM incorporates user-driven capabilities with a proven, rapid response capability to develop and support the model meeting evolving user needs. Teledyne Brown Engineering is the original and continuous developer of EADSIM, first deployed in 1989 [2].

REFERENCES

1. About EADSIM. See https://www.tbe.com/missionsystems/eadsim
2. USASMDC Public Affairs Office, "US Army Fact Sheet on Extended Air Defense Simulation". See https://www.smdc.army.mil/Portals/38/Documents/Publications/Fact_Sheets/EADSIM.pdf.

7 Overview of Synthetic Theater Operations Research Model (STORM)

Larry B. Rainey

NOTE TO READER

This chapter covers new and emerging technologies in the field of space warfare analysis. At the time of publication, there is limited information or accessible literature on this topic. As such, the chapter has been compiled by the editor from publicly available articles, fact sheets, and other online sources [1].

All sources in this chapter are in the public domain or reprinted with permission.

As stated in Chapter 1, I personally have used such tools as Systems Tool Kit (STK), Satellite Orbit Analysis Program (SOAP), System Effectiveness Analysis Simulation (SEAS), Extended Air Defense Simulation (EADSIM), and Synthetic Theater Operations Research Model (STORM) for the sake of conducting space system mission analysis. The first four tools are mission-level types of tools as addressed in Chapter 1. Advanced Framework for Simulation, Integration, and Modeling (AFSIM) and EADSIM are both mission- and engagement-type tools as also addressed in Chapter 1. The value added of STORM is that it is a campaign-level tool.

STORM falls under the International Traffic of Arms Regulations. Therefore, there is no publicly available information (i.e., fact sheet) concerning this chapter from the Department of Defense or otherwise. This chapter instead includes a naval post-graduate school thesis that addresses STORM. The core of the thesis is provided below.

* * * *

7.1 OVERVIEW OF STORM

To facilitate a basic understanding of STORM and a baseline for future research in the SEED Center, this chapter provides a detailed description of how STORM works, which includes its characteristics and framework, input and output, and how the U.S. Navy uses it. STORM is a stochastic, closed-form analytical simulation of air, space, ground, and maritime planning and execution. It is a campaign-level simulation designed to help decision-makers evaluate military strategy and capabilities in a theater of operations.

DOI: 10.1201/9781003321811-9

7.1.1 STOCHASTIC SIMULATION

In short, so-called mathematical, factors never find a firm basis in military calculations. From the very start there is an interplay of possibilities, probabilities, good luck
and bad that weaves its way throughout the length and breadth of the tapestry.

(Clausewitz, 1832, p. 86)

STORM is a complex, high-dimensional, stochastic, campaign-level simulation that
replaced its deterministic predecessor (a model known as ITEM) at OPNAV N81.
As Clausewitz indicates in the preceding quote, the nature of combat, along with
fundamental mathematical principles, implies that most combat simulations should
be stochastic because of combat's inherent randomness (Lucas, 2000). For instance,
a basic, deterministic Lanchester equation has one set of inputs and, therefore, only
provides one output. In stochastic modeling, the same set of inputs provides a range
of outcomes when the random seed(s) are changed for each replication.

7.1.1.1 Arguments for a Deterministic Combat Model

A common argument for using a deterministic model is that "a good point estimate is
sufficient for my purposes" (Lucas, 2000, p. 10). The point estimate, however, is frequently biased and does not provide information about variability (Lucas, 2000). Many
times, a decision-maker might know the outcome is in his favor; on average, however,
there is a significant chance that the outcome will be unfavorable. For example, if the
probability of failure is 1 percent on a five million dollar project, the decision-maker
might be willing to accept that risk. However, what if there is an 8 percent chance of
failure: Is the decision going to be different? Other arguments for deterministic models
can be similarly contested (Lucas, 2000). In a meeting with the developers of STORM,
they alluded to the fact that a deterministic model is just an intuition-confirmation
device with knobs that can be changed to get any result the analyst wants. In other
words, since there is no variability, one can adjust the input levels to get any result one
likes. This is not always the case with a stochastic simulation, since there may exist
chances for an unlikely result—even with the inputs set at generally favorable levels.

7.1.1.2 Arguments for a Stochastic Model

As stated previously, combat is inherently stochastic. Many uncertainties arise in
combat, such as outcome, process, future, and decision uncertainty (Lucas, 2000).
These factors are impossible to determine exactly. No battle fought will ever be
exactly the same as another because of changing technologies, terrain, strategy, and
human variability. As a result, the factors must be varied over a probability distribution to provide robust results of statistical significance.

The stochastic nature of STORM comes from the numerous data input parameters specified from 12 common probability distributions—including, but not limited to, the normal, binomial, and uniform. For example, if a ship has a damage and
repair capability, the amount of time it takes for the ship to be repaired may be pulled
from a uniform distribution, with a mean of three hours and a standard deviation of
six hours. Random variability exists throughout STORM, including in other areas
such as the probability of a hit or the probability of an intercept.

7.1.2   STORM—A CONSTRUCTIVE SIMULATION

STORM is a constructive simulation. A constructive model involves simulated entities, operating simulated systems, making decisions, and interacting. Real people
prescribe the decision-making logic to such simulations, but are not involved in
determining the outcomes once the inputs are provided (Department of Defense,
2010). In STORM's case, at OPNAV N81, analysts select input data, build a scenario— including the development of concepts of operations (CONOPS)—and execute the simulation. The simulation runs without a human in the loop and the analyst
receives the output data at the conclusion of the run (or set of runs, consisting of
many replications).

7.1.3   STORM – CAMPAIGN SIMULATION

STORM is a campaign-level simulation. A campaign is a series of related major
operations aimed at achieving strategic and operational objectives within a given
time and space (Department of Defense, 2014). Department of Defense (DoD)
simulations are normally classified into the following four categories: engineering, engagement, mission, and campaign. The spectrum of differences is wide.
A campaign model is one which is used to determine, for example, the best mix
of "blue" forces to battle "red" forces by focusing on order of battle and broad
probabilities of kill (Hawley & Blauwkamp, 2010). An engineering simulation,
which is at the other end of the spectrum in terms of detail, might only model a
certain weapon system's components and interactions. An example of an engineering model would be exploring the relationship of the weight of a bomb and
the range of an aircraft. As the weight of the bomb increases, the range of the
aircraft decreases. It might be ideal to include the level of detail in an engineering simulation in all simulations; however, at the campaign level, it is virtually
impossible to represent that much information for every entity. The result would
be an extremely long run-time and large amounts of memory required in order to
get a single run. This, of course, is not conducive to the timeline in which current
staff at OPNAV N81 operates.

   A theater in the context of military applications is the geographical area for which
a commander of a geographic combatant command has been assigned responsibility
(Department of Defense, 2014). STORM is typically utilized to simulate a single
theater or combatant commander (COCOM), in a time frame of weeks to months,
with a goal of completing operational objectives.

7.1.4   STORM CONCEPTUAL MODEL

STORM was originally developed as a campaign simulation by the Air Force. The
interesting approach in design that the developers took, however, was to ensure that
the model adapted an approach inclusive to definition, design, and development. The
end goal was a simulation that can be an interservice tool. Such a tool might reduce
the 10 "modeling wars" throughout the DoD. In addition, STORM was not hard-
wired to include only current day issues, but, instead, has the capability to include

the evolving environment of doctrine and operational concepts at low cost. This will maximize STORM's use over an operational life of perhaps 20 or more years.

7.1.4.1 Common Analytical Simulation Architecture

To maintain the flexibility desired, STORM employs the common analytical simulation architecture (CASA) decoupling components and applications, which endows modularity and segmentation, and insulates the system from local changes within individual segments. This enables STORM to operate with three different relational databases: Mini Structured Query Language (MSQL), Microsoft Office, and Oracle. In addition, this architecture enabled a database switch to be completed in less than two human days of programming effort (Group W, 2012a).

7.1.4.2 Storm's Logical Design

STORM models military operations from the real world with five classes: command and control (C2) manager, asset, intelligence manager, environment, and interaction manager. The flow concept of these classes can be seen in Figure 7.1.

7.1.4.3 Assets

Assets in STORM are entities that act or are acted upon. Therefore, they have the ability to complete activities like move, attack, conduct surveillance, or be killed. Examples of three different types of assets can be seen in Table 7.1.

Assets are tasked by the C2 manager and receive perceptions from the intelligence manager, weather changes from the environment manager, and state changes from the interaction manager. They also communicate status reports and state changes to the intelligence manager and interaction manager, respectively. Each type of asset has explicit information embedded in it such as mobility, location, and intelligence, surveillance, and reconnaissance (ISR) characteristics.

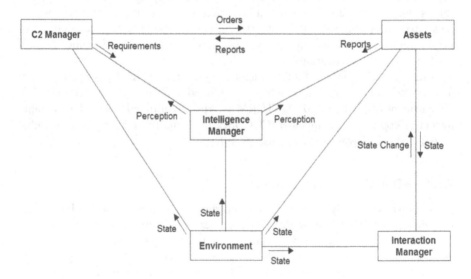

FIGURE 7.1   STORM's conceptual model (from Group W, 2012a).

TABLE 7.1
Asset Examples from STORM

Surface Asset	Air Asset	Orbital Asset
Armored Units	Aircraft	Satellites
Ships	Squadrons	Space-based platforms
Airbases	Munitions	
Logistics Nodes	Unmanned Aerial Vehicle (UAV)	

7.1.4.4   Environment

The purpose of the environment class is to model real-world environmental conditions such as time of day, weather, and terrain. The environment in which an asset is operating affects its capabilities. For example, environmental factors, such as high sea state, darkness, and dense fog, may affect the ISR capabilities of numerous assets.

7.1.4.5   Interactions

Interactions take place when two or more assets have the opportunity to affect one another. The three types of interaction managers are the motion managers, adjudication managers, and support managers. The motion manager is responsible for the movement of assets in response to their tasking and battle space dynamics, subject to resource and environmental constraints. The adjudication manager provides the result of engagements between two or more assets in combat, sensing, or communication missions. The support manager enables the movement and interaction of assets, subject to resource and environmental constraints, such as airbase operations (Group W, 2012a).

7.1.4.6   Intelligence

The intelligence manager provides perceptions to the C2 manager and assets. Information is gathered by different ISR platforms and analyzed to provide information like targeting data. Intelligence is not always correct—and, therefore, can lead to bad targeting data.

7.1.4.7   Command and Control

The command and control (C2) manager tasks and receives reports from assets. In addition, requests for intelligence are sent to the intelligence manager, and perceptions are sent from the intelligence manager back to the C2 manager. The objective of the C2 manager is to coordinate asset behaviors to meet operational and strategic goals. The decision-making process is modeled using optimization techniques and other algorithms.

7.1.5   Input Files

The information that populates the above-mentioned classes are contained in input files. The PUNIC21 scenario utilized in this thesis has over 100 input files. The files are available to view through STORM Front, which comes with the standard

```
 1  /*************************************************************
 2
 3     $Id: navalcommand.dat,v 1.1 2012/01/18 23:48:05 rlennox Exp $
 4
 5     File:   data/baseline_test/punic21/navalcommand.dat
 6     Rev:    1-1
 7     Date:   01/18/2012
 8
 9     Developed for the U.S. Government under contract(s):
10        N00178-04-D-4119
11
12     Classification: UNCLASSIFIED
13
14  *************************************************************
15
16     Define naval commands.
17
18  *************************************************************
19
20     Modification Log:
21
22     1-1:    01/18/2012  lennox/lburdette
23          : Initial implementation of Punic21 scenario
24          : for Detached Units Task 23DI0608.
25
26  *************************************************************/
27
28
29  Begin Naval_Command_File
30
31  Begin Naval_Command_List
32
33  ID: "Carthage Naval Command" {
34     Side:       "Allied Coalition"
35     ID: "Carthage Naval Fleet" {
36     }
37     ID: "Anglo Republic Naval Fleet" {
38     }
39     ID: "Blue Submarine Fleet" {
40     }
41  }
42
43  ID: "SWEMP Naval Command" {
44     Side:       "Red"
45     ID: "Western SWEMP Naval Fleet" {
46     }
47     ID: "Eastern SWEMP Naval Fleet" {
48     }
49     ID: "Red Submarine Fleet" {
50     }
51  }
```

FIGURE 7.2   Input file example for naval command.

STORM installation. It is a tedious job to understand what each file contains. An example of an input file can be seen in Figure 7.2.

The file designates the naval commanders for Allied and Red forces.

7.1.6 OUTPUT DATA IN STORM

A majority of the information from a replication can be found in the dbase.out and debug.out files. The dbase.out files contain raw data and must be processed through a relational database before any analysis can be done. The processing of data takes place in the data warehouse that is built into STORM. Once the data are loaded into the data warehouse, STORM contains three analyst tools (the map tool, the graph tool, and the report tool) designed to analyze and view the data. Although these tools are very easy to use and provide a quick way to look at different metrics, they are lacking in some desired capabilities. For example, not all output data are available to view in these tools. A programmer has the ability to write scripts in order to gain access to the information not included, but the typical analyst will usually require some additional training to accomplish this task.

7.1.6.1 Map Tool

The purpose of the map tool is to visually explore interactions taking place between different assets over time in a geographic region. The user is able to choose the time frame, how quickly the visualization of the simulation appears, and the geography and assets that they would like to be displayed. A screen shot of the map tool can be seen in Figure 7.3.

7.1.6.2 Graph Tool

The graph tool is designed to quickly and easily pull and graph user selected data from the data warehouse. Not all data are available in the graph tool, but there is a sufficient amount to gain insight rapidly on key output metrics of the simulation. An example of a graph tool output can be seen in Figure 7.4, which reveals the number of ships remaining for Blue and Red forces at the end of each day in the 20-day simulation.

7.1.6.3 Report Tool

The report tool function in the study tools of STORM allows the user to collect and organize specific data that can be further exported to conduct additional analysis. There is a wide selection of data available to collect in the report tool; however, like the graph tool, it is not inclusive of all output data from a simulation run. A nice

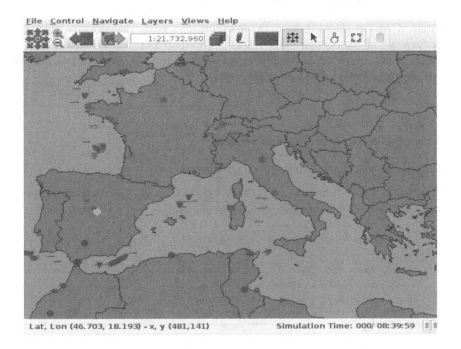

FIGURE 7.3 Screenshot of the map tool in STORM.

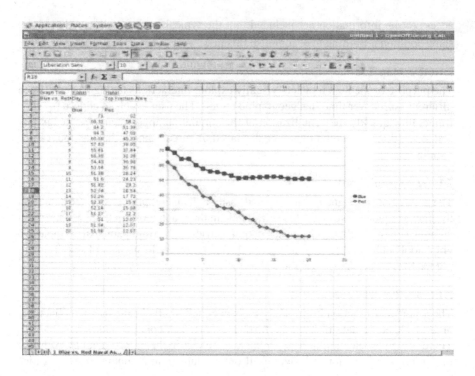

FIGURE 7.4   Screenshot of the Graph Tool in STORM.

feature of the report tool is the ability to export some of the data directly into a .csv file for analysis. An example of the output of the report tool can be seen in Table 7.2. This reflects a user-selected criterion to see which ships in the Blue Atlantic Surface Action Group (SAG) killed Red cruisers on Day 2.

TABLE 7.2
Screenshot of the Report Tool in STORM

Daily Maritime Kills (Rep Matrix)

Day	Killer Side	Killer Asset	Killer Naval Unit	Victim Side	Victim Target Type	1
2	Allied Coalition	Anglo Republic CSG S Cruiser 1	Blue Atlantic SAG	Red	Red Cruiser	1
2	Allied Coalition	Anglo Republic CSG S Cruiser 2	Blue Atlantic SAG	Red	Red Cruiser	1
2	Allied Coalition	Anglo Republic CSG S Destroyer 1	Blue Atlantic SAG	Red	Red Cruiser	1
2	Allied Coalition	Anglo Republic CSG S Destroyer 2	Blue Atlantic SAG	Red	Red Cruiser	1

7.1.7   Punic21 Scenario in STORM

For STORM to execute a run, a scenario is needed as input. The STORM installation comes with two unclassified baseline scenarios. This thesis focuses on one
of these scenarios, known as Punic21, in particular, due to its strong maritime
focus. This section describes the order of battle, geography, and phases of the war
to provide the reader with a basic understanding of how the scenario plays out.
The Blue forces consist of two allied nations known as the Anglo Republic and
Carthage. The Red forces are made up of the Swiss Empire (SWEMP). Tensions
have recently increased between Carthage and the SWEMP due to SWEMPs goal
of expansion. SWEMP secretly lays mines in the vicinity of Gibraltar to slow
Carthage's attempt to resupply the Anglo forces arriving in Spain. SWEMP forces
initiate attacks against Anglo naval forces and Integrated Air Defense Systems
(IADS).

7.1.7.1   Order of Battle

Based on the premise that this scenario is largely a naval campaign, the order of
battle includes the Blue and Red naval and air forces. For a campaign that centered
on land operations, the order of battle would include land forces such as army divisions, tanks, and artillery.

7.1.7.1.1   Naval Assets

The force structure of the naval assets can be seen in Table 7.3. The Blue forces have
an additional carrier, mine warfare ships, and an amphibious capability. The Red
forces have a few more destroyers and cruisers.

TABLE 7.3
Naval Order of Battle

	Blue Forces	Red Forces
Cruiser (CG)	8	14
Destroyer (DDG)	24	29
Nuclear Powered Aircraft Carrier (CVN)	3	2
Submarine Nuclear (SSN)	10	11
Guided Missile Submarine, Nuclear Powered (SSGN)	1	0
Mine Warfare Ship (MIW)	2	0
Landing Craft Air Cushion (LCAC)	3	0
Combat Logistics Force Ship (CLF)	11	3
OILER	6	3
Landing Helicopter Dock (LHD)	3	0
TOTAL	71	62

TABLE 7.4

**Air Order of Battle. Multi-role Fighter (MRF), Navy (N), Marines (M),
Early Warning (EW), Airborne Early Warning (AEW), Intelligence-
Surveillance-Reconnaissance (ISR), Unmanned Aerial Vehicle (UAV)**

	Blue Naval	Blue Air Force	Red Naval	Red Air Force
MRF—N	120	0	100	0
MRF—M	40	0	0	0
MRF—Tanker	15	0	0	0
MRF—EW	15	12	0	10
MRF	0	138	0	144
Fighter	0	70	0	64
Vertical Assault	40	0	0	0
AEW	9	12	3	10
MPA	12	0	8	0
Bomber	0	32	0	32
Tanker	0	36	0	0
UAV (ISR)	0	16	0	16
Airlift	0	24	0	24

7.1.7.1.2   Air Assets

The Blue forces have a slightly larger air capability, with additional multi-role
fighters (MRFs), fighters, and helicopters. The breakdown of the air forces can be
seen in Table 7.4.

7.1.7.2   Geography

The area of conflict is located in the Mediterranean Sea, the Bay of Biscay, and
the English Channel. The land geography in the scenario is Northwest Africa and
Western Europe. Figure 7.5 reflects a geographical outline with the location of
current military forces.

7.1.7.3   Phases of the Campaign

The scenario is separated into four phases: the Battle of the Atlantic, the Battle
of the Mediterranean, the fight for Spain, and the fight for Italy. These phases
overlap, but generally take place in the order listed above. Although no input
variables are changed, metrics that relate to these events are analyzed through
the rest of this thesis and take place at different times, due to the stochasticity
of STORM [2].

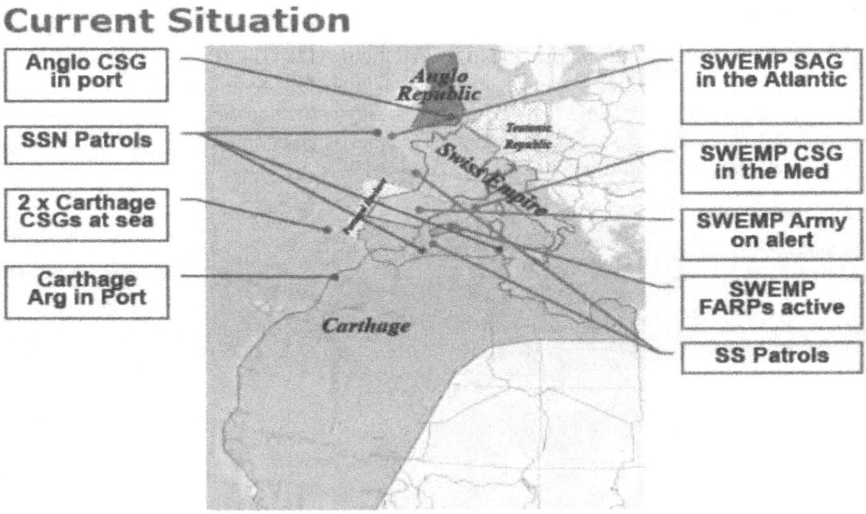

Current Situation

Anglo CSG in port	SWEMP SAG in the Atlantic
SSN Patrols	SWEMP CSG in the Med
2 x Carthage CSGs at sea	SWEMP Army on alert
Carthage Arg in Port	SWEMP FARPs active
	SS Patrols

FIGURE 7.5   Current Blue and Red force layout in geographical perspective (STORM).

* * * *

7.2 CONCLUSION

STORM is the primary campaign analysis tool used by the Office of the Chief of Naval Operations, Assessment Division OPNAV N81 and other DoD organizations to aid in providing analysis to top-level officials on force structures, operational concepts, and military capabilities. This thesis describes how STORM works, analyzes the variability associated with many replications, and evaluates the trade-offs between the expected number of replications and the precision and probability of coverage of confidence intervals. The results of this research provided OPNAV 81 with the ability to capitalize on STORM's full potential on a timeline conducive to its high-paced environment. The distribution of outcomes is examined via standard statistical techniques for multiple metrics. All metrics appear to have sufficient variability, which is critical in modeling the combat environment. The trade-off for confidence intervals between the expected number of replications, precision, and the probability of coverage is very important. If a more precise solution and a higher probability of coverage are required, more replications are generally needed. This relationship is explored and a framework is provided to conduct this analysis on simulation output data.

REFERENCES

1. Abstract found at https://apps.dtic.mil/sti/citations/ADA621426
2. Master's thesis entitled Capturing the Full Potential of the Synthetic Theater Operations Research Model (STORM) by Christian N. Seymour, September 2014. Overview of STORM located at pages 7 through 20. This thesis can be found at https://calhoun.nps. edu/handle/10945/44000

Section III

Space Wargaming Tools

8 Space Wargaming

Larry B. Rainey

NOTE TO READER

This chapter covers new and emerging technologies in the field of space warfare analysis. At the time of publication, there is limited information or accessible literature on this topic. As such, the chapter has been compiled by the editor from publicly available articles, fact sheets, and other online sources. All sources in this chapter are in the public domain or reprinted with permission.

The first section of this chapter features various articles as well as U.S. government and military fact sheets and press releases detailing the development and history of the Space Warfighting Analysis Center (SWAC). The second section includes a military fact sheet detailing the functions of the Space Wargaming Analysis Tool (SWAT).

* * * * *

8.1 SPACE WARFIGHTING ANALYSIS CENTER (SWAC)

8.1.1 PRIOR TO SWAC

8.1.1.1 Space Warfare Center

The mission of the original Space Warfare Center (SWC) was to advance America's space capabilities and employment concepts through tactics development, testing, analysis, and training programs. The genesis of the SWC dates back to Desert Storm, during which combat operations relied on space support more than any past conflict. However, an analysis of these operations revealed several shortfalls in the United States' ability to take advantage of all the capabilities space has to offer. To remedy this problem, in the fall of 1992, a blue-ribbon panel on space recommended the establishment of an SWC to examine the capabilities of space-based assets versus the actual enhancements gained from them.

The primary purpose of the SWC was to develop and test concepts, applications, and procedures that enable the warfighter to more fully utilize the unique capabilities of these space-based assets. The SWC is tasked with "operationalizing" space, making its use timely and routine to the warfighter. The aim is to exploit these Department of Defense (DoD), civil, and commercial assets in order to continue to provide U.S. forces with a definitive edge in future combat operations.

General Charles Horner, past commander of Air Force Space Command, officially dedicated the SWC on December 8, 1993. As the SWC grew, two testing squadrons were added in 1995 and the innovative Space Battlelab was dedicated in 1997.

The mission of the 17th Test Squadron (TS) is to enhance Air Force Space Command (AFSPC's) support to the warfighter through testing and evaluation of space forces. The 17th TS is responsible for the planning, execution, and reporting for all follow-up operational test and evaluation of Air Force space forces.

The mission of the 576th flight test squadron is to manage and conduct the Joint Chief of Staff (JCS)-directed Intercontinental Ballistics Missile Follow-on Operational Test and Evaluation program. The squadron has the sole responsibility for collecting, analyzing, and reporting the data required to verify the accuracy and reliability of our deployed ICBMs.

The mission of the Space Battlelab is to identify innovative space operations and logistics concepts and rapidly measure their potential for advancing the Air Force's core competencies in joint warfighting. The Space Battlelab uses field ingenuity, modeling and simulation, and actual employment of exploratory capabilities in operational environments to accomplish the mission.

The objectives of the SWC are

- Advance the Air Force's core competencies;
- Ensure space and missile systems meet the needs of the warfighter;
- Integration of present and future space systems into military operations;
- Train warfighters, space systems operators, and space support teams;
- Develop, validate, and document tactics for the employment of space systems and capabilities; and
- Visual modeling and simulation of the effect of space systems on the battlespace [1].

* * * * *

8.1.1.2   Longevity for the Concept of a Space Warfare Center

As stated above, the original SWC was stood up on December 8, 1993, under the dedication of then General Charles Horner Commander of then Air Force Space Command. The SWC was redesignated as the Space Innovation and Development Center on March 1, 2006 [2].

On April 1, 2013, Air Force Space Command inactivated the Space Innovation and Development Center and its subordinate units were reassigned. The base's host unit, the 50th Space Wing, gained the 3rd Space Experimental Squadron from the inactivation while remaining units and missions transferred to units of the U.S. Air Force Warfare Center [3].

No really great idea stays unattended to. The establishment of the SWAC was ordered by Chief of Space Operations John W. Raymond [4]. Originally planned as Space Warfighting Integration Center, Vice Chief of Space Operations David D. Thompson was tasked to focus on its establishment upon taking office [5]. Raymond approved the organization design of SWAC on March 8, 2021 [6].

* * * * *

8.1.1.3   Space Warfighting Analysis Center (SWAC)

"Space Warfighting Analysis Center has unique functions, official says."

Vice-Chief of Space Operations Gen. David Thompson, a past Potomac Officers Club event speaker, said there is no overlap in the functions of the SWAC and the Space Security and Defense Program (SSDP).

Thompson pointed out the unique value offered by the SWAC during an event hosted by the Mitchell Institute amid backlash from lawmakers, who denied the nascent organization a $37 million budget due to perceived duplication of functions with SSDP.

He explained that the U.S. Space Force's SWAC handles the force design and configuration of the nation's myriad satellite fleets, while the SSDP, which is jointly managed by the DoD and the Intelligence Community, focuses on the "protect and defend" mission for U.S. military and spy satellites, Breaking Defense reported Wednesday. Force design assessment is facilitated by conducting analysis, modeling and simulation, wargaming, and experimentation to create operational concepts.

Thompson said SSDP is not charted to make decisions regarding the design of the nation's space data relay and positioning, navigation, and timing enterprises.

Another cause of concern for lawmakers is the SWAC's overlap with the National Reconnaissance Office (NRO), particularly the area of buying commercial intelligence, surveillance, and reconnaissance data.

The Space Force addresses the concern, saying it has begun contributing to the NRO's mission of providing airborne ISR data to offer more resilient coverage for future fights with peer competitors.

According to Thompson, there will be plenty of work to distribute between the Space Force and the IC. SWAC was launched in line with the CSO Planning Guidance released in November by Space Force head, Gen. Jay Raymond. The organization is working on finalizing an architecture study for the composition of a U.S. missile warning satellite network [7].

8.2   SPACE WARFARE ANALYSIS TOOL (SWAT)

SWAT enables the rapid and dynamic creation and execution of multiple platforms for wargaming courses of action. SWAT generates critical data that can be used to inform commanders and decisionmakers regarding space concepts, capabilities, concept of operations, and tactics; and techniques and procedures in environments with and without space-based capabilities. SWAT supports space, air, and ground maneuver (Red and Blue) forces in a wargaming environment, while providing data collection and reduction in real time, allowing the user to get an understanding of the impacts of planned and injected events and platforms. SWAT is not focused on system engineering of detailed operational analysis but for quick looks. SWAT trades off fidelity for ease of use. It can be tailored for future capabilities, without the need of full system definition to run. The 3D map gives commanders and leaders a unique understanding of the following contributions, benefits, and limitations of space—both Red and Blue:

- Quick scenario generation/editing/execution
- Intelligence, surveillance, and reconnaissance collection planning and coverage visualization and effectiveness testing
- Defense laydown effectiveness insights (sensor/launcher emplacement)
- Insight on the battlefield effectiveness of organizational designs and systems
- Communication architecture planning/visualization (sensor to shooter)
- Developing augmented reality view and interactions for users

8.2.1 SIMULATING THE BATTLEFIELD

SWAT is easy to use, generates quick scenarios, and enables high-level analysis of single or multiple platforms supporting space, air, and ground maneuvers (Red and Blue) forces in a wargaming environment, while facilitating quick looks of courses of action.

Warfighters make and revise strategies on a continual basis, resulting in the need for a tool that develops multidomain scenarios to provide situational understanding of the potential impacts of strategic and tactical decisions.

SWAT simulates the battlefield in a rapidly deployable quick-scenario generation and execution tool that enables high-level analysis of single or multiple platforms supporting the armed forces in a wargaming environment. SWAT is used to gain a real-time understanding of the effects of its actions against an adversary. This government-off-the shelf tool is available to all DoD organizations wanting to visualize a space-enabled wargaming environment.

SWAT's primary function is to generate Red-team versus Blue-team wargames, where two sides battle each other virtually with planned or dynamically injected interactions and the effects of those actions play out against each other. SWAT can enable the dynamic deployment of battle assets—divisions, brigades, battalions, companies, platoons, and fire teams—in real time or faster than real time, providing further insight into potential scenarios. These scenarios can include uncommon events, such as degradation or jamming of communications, which force users to adjust strategies on the fly.

SWAT imports satellite flight paths and satellite constellations so that all domains, including space, are included in the wargaming analysis. A comprehensive understanding of all outcomes requires a comprehensive input of scenarios, and SWAT achieves just that.

The program is accessible on a laptop and tablet in both Windows and Linux.

SWAT provides a 3D-game view with augmented reality being developed. In the future, augmented reality headsets and peripherals will provide a virtual sand table to enable easy collaboration and communication.

The user interfaces in both augmented reality and desktop views are based on National Aeronautics and Space Administration's Worldwind Map Engine data. This is an open-source virtual globe that allows developers to quickly create interactive and accurate visualizations of the Earth. In addition, SWAT can use compressed ARC digitized.

Highlights
- Develop "base case" scenarios consisting of space, air, and ground entities with planned paths.
- Utilize Satellite Orbit Analysis Program file data for satellite constellation paths.
- Scenarios include planned events entered into the base case scenario.
- WorldWind 3D Terrain Elevation Engine, Digital Terrain Elevation Data levels 0–2, and compressed ARC digitized raster graphics.
- Red versus Blue with multiple user views of the battle.
- Interactive mode that allows for dynamic injection of entities and event in real time. Events would include (but not limited to) degradation of communications, jamming, spoofing, and damage.

- SWAT supports a faster than real-time function but can be paused, saved, and edited at any point in the scenario.
- Detailed events describing space, air, and ground force states at different points of the scenario.
- Supports an initial capability of an augmented reality view of battle [8].

* * * * *

SUMMARY

This chapter addressed the subject of space wargaming. Eight articles comprised this chapter. The first related to the standup of the SWC at Schiever Air Force Base, Colorado. The second addressed the redesignation of SWC to the Space Innovation and Development Center (SIDC). The third pertained to the deactivation of SIDC and the distribution of its mission to two other organizations. Articles 4, 5, and 6 addressed the standup of the SWAC in Colorado Springs, CO. Article 7 went into detail to describe the mission of SWAC. Article 8, the final article, described a specific tool called SWAT.

Chapters 9 and 10 will address the details of two specific space wargaming tools. They are Space Warfare Analysis Tools/Space Attack Warning and Space and Information Analysis Model.

REFERENCES

1. "Space Warfare Center," GlobalSecurity.org. See https://www.globalsecurity.org/space/agency/swc.htm.
2. Robertson, Patsy. *Space Innovation and Development Center (AFSPC) Fact Sheet.* Air Force Space Command, March 1, 2006.
3. Saunders, Randolph. "Schriever: A brief history," Air Force Space Command (Archived), September 8, 2016. See https://www.afspc.af.mil/News/Features/Display/Article/947127/schriever-a-brief-history/.
4. "Chief of Space Operations Planning Guidance 1st Chief of Space Operations." US Space Force. See https://media.defense.gov/2020/Nov/09/2002531998/-1/-1/0/CSO%20PLANNING%20GUIDANCE.PDF.
5. Kirby, Lynn. "First-ever Vice CSO joins U.S. Space Force." Secretary of the Air Force Public Affairs, October 4, 2020. See https://www.spaceforce.mil/News/Article/2371102/first-ever-vice-cso-joins-us-space-force.
6. Space Operations Command (USSF). See https://www.afhra.af.mil/About-Us/Fact-Sheets/Display/Article/2886917/space-operation-command-ussf/
7. Rosenberg, Adrian. "Space Warfighting Analysis Center Has Unique Functions, Officials Says." Potomac Officer's Club, July 29, 2021. See https://potomacofficersclub.com/space-warfighting-analysis-center-has-unique-functions-official-says/0.
8. "SPACE WARGAMING ANALYSIS TOOL (SWAT)". USASMDC Public Affairs Office. See https://www.smdc.army.mil/Portals/38/Documents/Publications/Fact_Sheets/SWAT.pdf.

9 Space Warfare Analysis Tools (SWAT) and Space Attack Warning (SAW)

Paul Szymanski

Part 1: Software Detailed Descriptions

9.1 SUMMARY

Paul Szymanski has 50 years of continuous experience in space control analyses and translating military concepts and methods from terrestrial warfare to new doctrines, strategies, and concepts for space. These combined experience bases have been integrated into the Space Warfare Analysis Tools (SWAT) software. The main technical thrusts for this SWAT development are listed in Table 9.1.

All SWAT functional capabilities can provide significant support to Space Situational Awareness (SSA), Space Domain Awareness (SDA), Space Predictive Battlespace Awareness (PBA), and Space Attack Warning (SAW) activities. SWAT automatic space object state change detection algorithms can show meaningful changes in orbits, size, optical, and RADAR signatures. These indicators can easily be viewed in the SWAT Choke Point map displays. Analyses of SWAT optical data show that this can significantly impact state change results. Attack indicators have been developed and incorporated into SWAT automatic red space course of action (COA) detection algorithms. Additional tools developed: an automatic space systems scenario generation tool and Microsoft Project 5,000-line space control scenario were designed in support of the automatic red space COA identification work. In addition, optical characteristics data for space objects were obtained from Maui and internet sources and placed in a Microsoft Access database of 138,000 detection records. Also, new definitions were developed for space control activities that were extrapolated from terrestrial doctrine, and are listed in the Glossary in the back of this chapter. Finally, SWAT participated in the Advanced Concept Event (ACE) exercise test run to further evolve user interfaces and techniques.

TABLE 9.1

Main SWAT Technical Thrusts

1. Automatic Space Object Change Detection
2. Automatic Detection of Attacks In Space
3. Space Choke Point Map Displays
4. Automatic Red Space Courses of Action (COA) Identification
5. Automatic Space Wargaming Scenario Generation

 DOI: 10.1201/9781003321811-12

9.2   IDENTIFICATION AND SIGNIFICANCE OF THE PROBLEM

9.2.1   GOALS

The fundamental goal for the activities described in this section is to support improved SSA, SDA, PBA, and SAW. Because the world has not yet experienced a full-out space war, it is difficult to assess what the likely conditions, battlefield tempo, strategies, and tactics would underlay a future space conflict. Nevertheless, it is likely that some potential adversary of the United States is currently devoting considerable resources to designing systems that can conduct surprise assaults on strategic space assets of US and allied countries. Due to the distant (up to 36,000 km and more) and unmanned nature of satellite systems, detection of these attacks would be difficult, and currently may only be of the "post-mortem" variety. Even with major resource allocations by nations towards intelligence gathering through all mediums, history is replete with examples of major surprise attacks that should have been detected, but were not (e.g., Pearl Harbor, Battle of the Bulge (in spite of 11,000 ultra message decryptions indicating buildup of major German forces for this attack), Yalu River in Korea, most Israeli-Arab conflicts). The ability to detect attacks in distant space can only be more difficult and less certain than these terrestrial examples. In addition, there are many historical examples of new weapon technologies that provided considerable advantages to their first user, and fundamentally changed the correlation of offense vs. defense in their respective theaters, at least for a period of time (catapult vs. Greek fortifications, cannon vs. castle walls, crossbow vs. shield, musket vs. body armor, tank vs. machine gun, shaped-charge vs. bunker, airplane vs. battleship, etc.). More than likely the side that first employs offensive weapons against space systems will "win" the space war, and unbalance US and allied use of space systems to support the terrestrial battlefield, at least over the short duration of any probable future conflict. Because of all the above uncertainties concerning detecting surprise attacks against space systems, considerable thought has to be applied towards SSA, space PBA and SAW algorithms, techniques, procedures, and databases. It is in the nation's interest to pursue multiple approaches to these issues, in order to cherry-pick the best of each, so as to insure against an uncertain world.

9.2.2   BENEFITS

With warfighter use of SWAT, requirements for space warfare can be more fully understood. Space objectives, strategies, tactics, checklists, and space conflict timelines would be readily available to the warfighter to better understand the space environment and threats. Operationally, this SWAT tool can provide an analytic "tripwire" in orbital space denoting when potentially threatening satellites can be designated with increased levels of threat awareness as they get too close to key Allied satellites, and they can be rank-ordered for attention by SSA assets. Key intelligence indicators can be tracked and SSA assets optimized for detection of the most probable adversary COA. Finally, this tool can help train warfighters in sensitizing them to which particular orbits should have the most attention placed for defense of critical space assets. Use of this tool may very well provide strategic warning of space attack, and thus early warning of possible military actions on the ground.

9.3 SWAT INTRODUCTION

9.3.1 TECHNOLOGY HISTORY

Concepts leading up to the invention of this tool's algorithms have a long history. In the 1980s, this author developed space principles of war that were used by the Joint Chiefs of Staff when developing plans for the formation of Air Force Space Command. After 48 years of space control analyses, the author has also developed a list of 5,000 different space-related military objectives. In addition, he has designed many space strategies, tactics, and procedures useful for space control concepts, a space NIIRS intelligence collection rating scale, and a Space Centers of Gravity Model (based on Col. John Warden's (Checkmate) 5-Ring COG Model). Previous analytic work was performed to develop extensive lists (~1,900 factors) of Space Intelligence Preparation of the Battlespace (IPB) indicators to determine if a space system is under attack. Prototype SWAT code implementing all of the above concepts and databases was developed in Microsoft Access to test the algorithms for space IPB and attack warning and assessment.

9.3.2 OVERALL INTEGRATED CONCEPT

Figure 9.1 illustrates the overriding algorithm that SWAT employs to support the SSA and SAW missions. This process starts with extensive data on existing satellite and space systems characteristics such as orbital elements, size, shape, RADAR

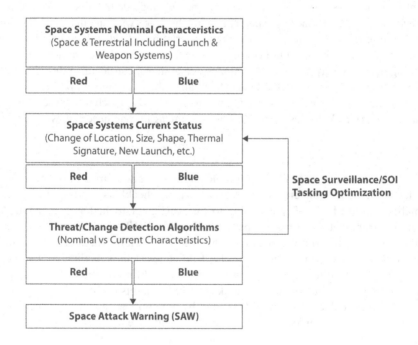

FIGURE 9.1   SSA & SAW integrated concept—Flow 1.

TABLE 9.2
SWAT Space Object Characteristics List

Physical Characteristics	Orbital Characteristics
• Radar Cross Section (RCS)	• Inclination
• Optical Cross Section	• Eccentricity
• Flashing or Not	• Mean Motion
• Flash Period	• Mean Anomaly
• Stabilization Type (Spinning or 3-Axis)	• RAN
• Object Shape (Sphere, Cylinder, Box)	• Argument of Perigee
• Length	• BStar
• Width	• 1st Mean
• Height	• 2nd Mean
• Mass	
• Spin Rate	
• Delta-V	
• Satellite Position (Geosynchronous)	
• Beginning of Life On-Board Power	
• Major COMM Antennas & COMM Signals	
• Major Optics On-Board	
• Retro Reflectors On-Board	

cross-section (RCS), optical cross-section, stabilization, size, etc. (see Table 9.2 for a complete list). These characteristics are derived over the long term, and can include historical data over many years. The AFRL Satellite Assessment Center's (SatAC) Satellite Information Database (SID) provides such a database [1]. This historical database can be considered the "steady state" condition of the space system, and includes the range of data values in which the space system can be considered operating within its "normal" bounds.

The second set of data for satellites, space objects, and terrestrial systems supporting space includes what these systems are currently doing. This is the current status of these space systems. This can be measured by space surveillance, SIGINT, and other intelligence collection methodologies.

The third step in the SWAT process is to use change detection algorithms to compare the historical steady-state nature of the space system to what its current state is. Because of the many thousands of space objects in existence today (including live and supposedly dead space objects), only automatic algorithms can identify those space objects that require further analysis and additional data collection/monitoring. This process can then identify those space objects that may be changing status (configuration, orbital maneuvering, or ejecting sub-satellites), and may be threatening other space systems. It should be noted that these processes and algorithms must also be applied to Red, Blue, Gray and "dead" space objects in order to accomplish a complete picture of what the current situation in space is. The change detection for the Blue space systems is required in order to support attack assessment and battle damage assessment (BDA).

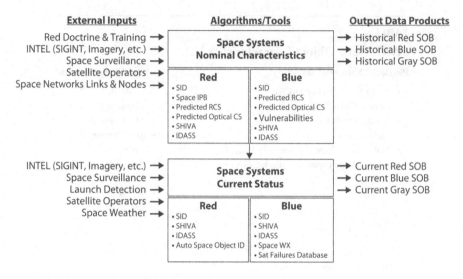

FIGURE 9.2   SSA & SAW integrated concept—Flow 2.

Figure 9.2 expands on the previous chart and lists the data inputs and product outputs of the first two steps of the SWAT algorithms processes. In the Red and Blue boxes, AFRL research programs are listed that database or assess intelligence data on space systems.

Figure 9.3 further details the SWAT change detection algorithms, and lists the data inputs and product outputs of the second two steps of the SWAT algorithms process. These process steps are where change detection and Satellite Attack Warning

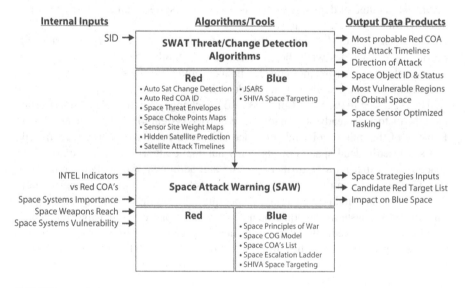

FIGURE 9.3   SSA & SAW integrated concept—Flow 3.

(SAW) are determined in SWAT. In the Red and Blue boxes, AFRL research programs are listed that process or assess intelligence data on space systems.

9.3.3 BACKGROUND

The activities delineated under this chapter are part of the overall architecture for the SWAT that address the fundamental SSA, SDA, SAW, and PBA questions listed in Table 9.3.

These space warfighting tools are based on the principal investigator's 49 years of continuous experience in space control analyses, with military concepts and methods from terrestrial warfare applied to new strategies for space. They form one of the first consistent structures for thinking about space warfare, and provide a starting point for research on space battle management and SSA requirements. They are also immediately applicable to operational use in a militarized space environment for developing strategy, tactics, and attack detection procedures. SWAT helps predict an adversary's intentions and next moves against space systems. Mr. Tom Behling, when he was the Deputy Undersecretary of Defense for Preparation and Warning, described SWAT as "Unique in this country, with no one else conducting this kind of far-thinking research in SSA" and "Critical to this nation's defense in space." Ultimately, the SWAT approach provides a "Unified Field Theory" for SSA and SAW.

SWAT tools and databases include 1,900 requirements for IPB and space attack indicators (Table 9.4); space NIIRS (National Imagery Interpretability Rating Scale) intelligence collection rating scale (Table 9.5); Space Principles of War; space strategies, tactics and procedures; Auto Space Object ID (98 percent accurate with size, RCS and orbital element data); Auto Red Space COA ID (Figure 9.4); Satellite State Change Detection (Figure 9.5); Space Threat Envelopes (most threatening regions of space) (Figure 9.6); Space Choke Point Maps (Figure 9.7); Most Probable Space Attack Time Maps (Figure 9.8); space Centers Of Gravity (Table 9.6); space conflict escalation ladder; 5,000 space-related military objectives (Table 9.7); Satellite Characteristics Database and requirements tracking software; automatic space scenario generation (Figure 9.9); and Space INTEL Tasking Prioritization algorithms. In addition, the Space Highest Information Value Assessment (SHIVA) software tool automatically ranks satellites and terrestrial space-related elements for their value in enabling the flow of worldwide critical military data supporting the battlefield, and can be used as a common ranking tool for both space and terrestrial targeting.

TABLE 9.3

Fundamental SSA /SDA/SAW /PBA Questions

1. Will Space Systems Be Under Attack In the Near Future?
2. Are Space Systems Currently Under Attack?
3. Who Is Attacking?
4. What is the Adversary Attack Strategy?
5. What Damage Has Been Caused?
6. What Is Optimal Blue Response?

TABLE 9.4

Example SSA Requirements Matrix (Partial Data)

Is the weapon system preparing/powering up for use?

INTEL Derived From	INTEL Requirements	INTEL Indicators	Resolution Requirements	Space NIIRS	Detection Means	Technologies
Basic Characterization	**Satellite Current Orientation Attitude**	Satellite Current Cross Section	1.2–2.5	4	Imagery or RADAR	Optical or RADAR
Basic Characterization	Satellite Has Changed Attitude From Spinning or 3-Axis Stability	Satellite Cross Section Change	2.5–4.5	3	Optical or RADAR Cross Section	Optical or RADAR
Detailed Characterization	Satellite Current Weapons Suite Pointing Direction	Satellite Weapons Suite Image	0.20–0.40	7	Imagery	Optical
Exquisite Characterization	Satellite Delta-V Remaining Capability	Satellite Telemetry Indicates Propulsion Tank Fluid Level	N/A	N/A	RF Signal Monitoring	RF Receivers
Exquisite Characterization	**Satellite Propulsion Tank Fluid Status**	Satellite Propulsion Tank Thermal Image	0.20–0.40	7	Imagery	Optical-IR
Exquisite Characterization		Satellite Telemetry Indicates Propulsion Tank Fluid Status	N/A	N/A	RF Signal Monitoring	RF Receivers
Exquisite Characterization		Satellite Propulsion Tank Thermal Image	0.20–0.40	7	Imagery	Optical-IR
Exquisite Characterization	Satellite Current On-Board Processor State	Satellite Telemetry Indicates On-Board Processor State	N/A	N/A	RF Signal Monitoring	RF Receivers
Exquisite Characterization	Satellite Propulsion Tank Internal Pressure	Satellite Telemetry Indicates Propulsion Tank Internal Pressure	N/A	N/A	RF Signal Monitoring	RF Receivers
Detailed Characterization	Satellite Current Detailed Thermal Signature	Satellite Thermal Image	0.20–0.40	7	Imagery	Optical-IR

TABLE 9.5
NIIRS Space Equivalents

NIIRS Rating	GRD (m)	Terrestrial Examples	Space Equivalent Examples
		National Imagery Interpretability Rating Scale (NIIRS)	
0		Interpretability of the imagery is precluded by obscuration, degradation, or very poor resolution	Satellite features in shadow
1	9	Detect the presence of aircraft dispersal parking areas.	Detect the presence of very large (e.g., International Space Station) space object.
2	4.5–9.0	Detect the presence of large (e.g., Boeing 737, 747, Airbus A-300, MD-80) aircraft.	Detect the presence of large (e.g., GEO Communications satellite) space object.
3	2.5–4.5	Detect medium-sized aircraft (e.g., F-15). Identify an ORBITA site on the basis of a 12-meter dish antenna normally mounted on a circular building.	Detect the presence of a medium (e.g., DMSP) space object.
4	1.2–2.5	Identify the wing configuration of small fighter aircraft (e.g., F-16). Detect large (e.g., greater than 10 meter diameter) environmental domes at an electronics facility.	Detect if large (e.g., TDRS) solar panel has deployed.
5	0.75–1.2	Distinguish between single-tail (e.g., F-16) and twin-tailed (e.g., F-15) fighters. Detect automobile in a parking lot. Identify the metal lattice structure of large (e.g., approximately 75 meter) radio relay towers.	Determine large (e.g., TDRS) solar panel design configuration. Determine satellite attitude/spin rate. Determine if satellite has broken up into large pieces.
6	0.40–0.75	Detect wing-mounted stores (i.e., ASM, bombs) protruding from the wings of large bombers (e.g., B-52). Identify the spare tire on a medium-sized truck.	Determine existence of medium-sized (TDRS SGL Antenna) satellite antennas.
7	0.20–0.40	Identify antenna dishes (less than 3 meters in diameter) on a radio relay tower. Identify individual 55-gallon drums. Detect small marine mammals (e.g., harbor seals) on sand/gravel beaches. Identify ports, ladders, vents on electronics vans. Identify individual rail ties.	Determine attitude of medium-sized (TDRS SGL Antenna) satellite antennas. Determine large area degradation of solar panel optical quality.
8	0.10–0.20	Identify the rivet lines on bomber aircraft. Detect horn-shaped and W-shaped antennas mounted atop BACKTRAP and BACKNET radars. Identify windshield wipers on a vehicle. Identify limbs (e.g., arms, legs) on an individual. Identify individual horizontal and vertical ribs on a radar antenna.	Determine medium-sized (TDRS SGL Antenna) satellite antenna damage.
9	<0.10	Identify screws and bolts on missile components. Detect individual spikes in railroad ties. Identify individual rungs on bulkhead mounted ladders. Identify vehicle registration numbers (VRN) on trucks.	Detect orbital thruster damage. Detect internal fuel reserves by IR means.

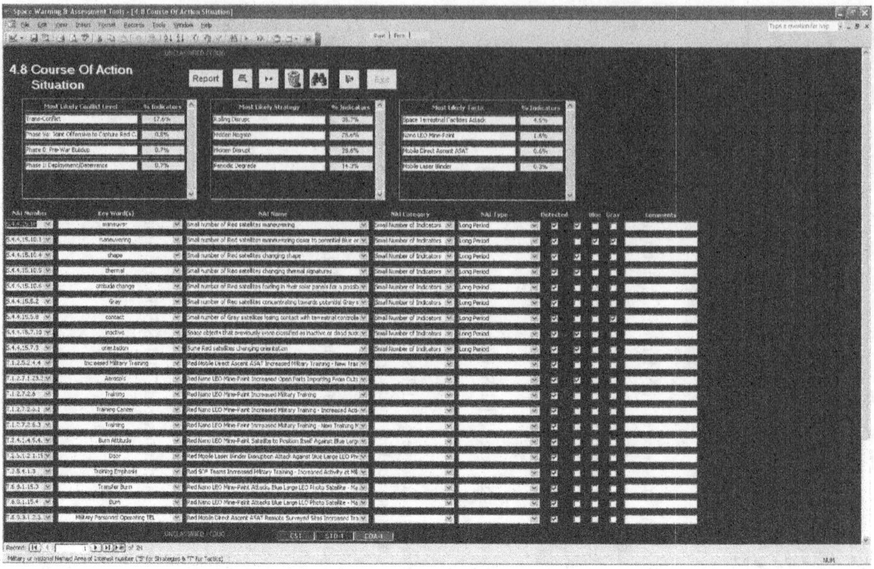

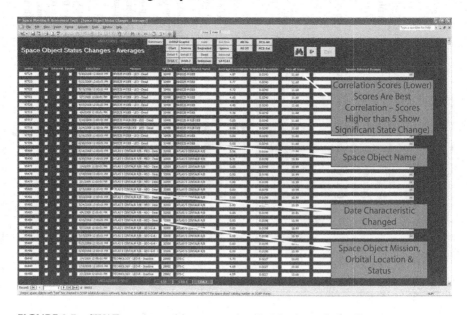

FIGURE 9.4   Example SWAT auto attack assessment.

9.3.4   SWAT Tools

Below lists the main features of tools and algorithms currently available or requiring additional development under SWAT that help space operators and analysts determine critical space control and Space Situational Awareness factors leading to informed decision making for space COAs.

FIGURE 9.5   SWAT auto state change example prioritized correlation list.

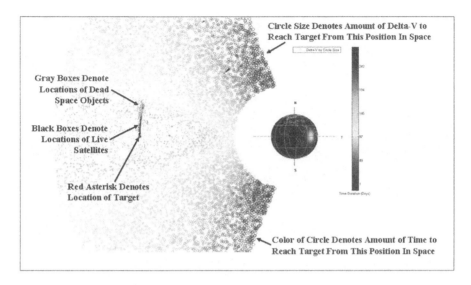

FIGURE 9.6   Example space threat envelope (low thrust maneuver).

9.3.4.1   Auto Space Object Mission Classification

SWAT software contains algorithms that automatically classify the mission of any unknown space object. This algorithm uses the Satellite Information Database (SID) [2] and the Space Power Analysis and Requirements Keystone Software (SPARKS) databases and compares the satellite characteristics and associated missions of these databases with any known characteristics of the unknown space object. Example

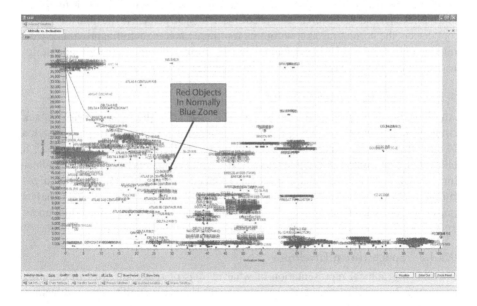

FIGURE 9.7   SWAT choke point map.

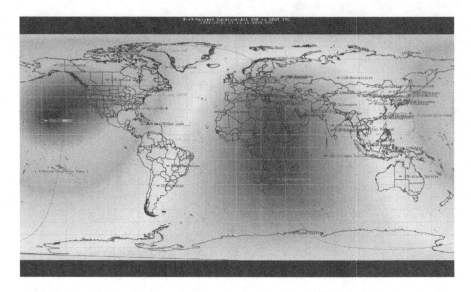

FIGURE 9.8   Example attack locations optimized for space surveillance.

satellite characteristics that are matched with the unknown space object include RCS (Figure 9.10), optical cross-section (Figure 9.11), orbital elements, stabilization type (spinning or 3-axis), object shape (sphere, cylinder, box), length, width, height, mass, and spin rate (see Table 9.2 for a complete listing of space object characteristics used in SWAT algorithms. The characteristics listed in red are those that are optical-related).

If the SWAT user knows, at a minimum, the RCS, size (through RADAR imaging) and orbital elements of the unknown space object, then 98 percent of the time the SWAT tool detects the correct mission within its top three choices (out of approximately 50 mission options). SWAT has even automatically corrected some of the mission designations in the NORAD satellite catalog and detected incorrect catalog numbers.

9.3.4.2   Auto Red Space COA ID

An algorithm was developed to automatically determine the most probable attack strategy that an adversary may be conducting against US and allied space systems (see Section 9.8 "Automatic Space Course of Action Detection Tool," page 152). It takes sometimes obscure intelligence indicators, and compares them to those indicators related to possible adversary strategies in space warfare. In addition, other algorithms and software calculations limit the range of attack possibilities that an analyst must consider when determining that high-value space assets may soon come under attack, or have actually already been attacked. These tools include the following.

9.3.4.3   Space Threat Envelopes

Provides a graphical representation of what regions of space are most threatening to a particular satellite, based on assumptions of adversary satellite size and Delta-v maneuver constraints (Figure 9.6).

TABLE 9.6
Example Space Centers of Gravity (Partial List)

Launch corridors
GEO belt sectors
Sun-Synchronous LEO orbits
GEO satellites changing orbital position
Space-related command centers / commanders / INTEL Centers
Space surveillance systems
Space technicians / scientists
Electric grid serving ground space facilities
Space design and manufacturing facilities
Leader's confidence in their new space technologies
Blue and Red side political will to start and continue a space war
Space-related decision cycle times (OODA loops)
Low delta-v/transit time points in space to reach High Value Targets
Points in space with high/low coverage from space surveillance assets
Regions of space and time with advantageous solar phase angles
Times of high solar storm activity
On-orbit spares or launch replenishment or ability to reconstitute space capability with terrestrial systems
Antipodal nodes 180 degrees from launch sites around the world
Manned launch (Shuttle, Space Station) of satellites
Initial satellite checkout after launch or orbital insertion
Periods of solar eclipse / low battery charge for satellites
Approach trajectories outside the field of regard of the target's on-board sensors
Approach trajectories when the Sun/Moon/Earth is in the background of a target's sensors
Approach trajectories outside normally employed orbits
Near a satellite's thrusters
Near a satellite's high power antennas
Just after loss of contact with adversary satellite ground controllers / space surveillance assets

9.3.4.4   Space Choke Point Maps
Provides a graphical representation of what regions of space (altitude vs inclination) that space objects concentrate in, whose boundaries are limited by Delta-v maneuver constraints (see Section 9.6 in "Part 2: Example Analyses," page 142).

9.3.4.5   Most Probable Attack Time Maps
These charts show where on the earth a potential satellite attack may occur. If one assumes that an attack will occur where the US has the fewest space surveillance assets to monitor potential attacking space objects, then the redder the map is, the fewer US space surveillance assets exist, geographic-wise. The bluest portions of the map are a geographic average of the locations of US space surveillance assets. For illustrative purposes, the redder portions of the map give the geographic average of SPOT ground data receiver and Tracking, Telemetry & Control (TT&C) stations (Figure 9.8).

TABLE 9.7

Example Space Objectives (Partial List)

Blind Blue capabilities to observe the terrestrial battlefield

Blind Blue capabilities to observe space from terrestrial sensors

Blind Blue capabilities to observe space from space-based sensors

Spoof Blue capabilities to observe the battlefield

Deny Blue ability to launch new satellites

Destroy some Blue space capability as a warning to Gray space systems support to Blue Wear down Blue Defensive Counter-Space capabilities by instigating multiple false alarm attacks

Attack Blue satellites before the start of the terrestrial conflict

Spoof Blue perceptions of Red space strengths

Conduct diplomatic offensive to restrict Blue ability to employ ASATs

Actively defend key launch corridors and orbits critical to Red conduct of war

Preposition Red space assets to maximize their effectiveness at the start of the conflict

Disrupt Blue command and control capabilities for space systems

Embargo Blue access to space systems

Prevent Blue ability to service or re-fuel on-orbit satellites

Develop propaganda campaign against Blue use of ASATs

Shape and delay Blue plans for space warfare

Deny Blue ability to achieve Space Situational Awareness

Disrupt Blue space attacks so they become uncoordinated

Constantly shift points of application of space control weapons to contuse adversary response

Herd Blue space communications paths to those that are more easily monitored by Red SIGINT assets

Attack key Blue space personnel and technicians

Disperse Red assets (maneuver satellites) just before launching first attack

9.3.4.6 Auto State Change Detection

This algorithm is similar to the automatic space object mission identification tool, only now it compares satellite orbits and characteristics with other satellites of the same mission, and with its own historical data. The satellites with the most significant changes are presented to the space warfare operator so that he can task space surveillance sensors to determine if anything suspicious is happening with this satellite or space object. This tool may also be able to automatically predict when a satellite is preparing to maneuver, re-configure to an anti-satellite (ASAT) mission, or is beginning to fail (see Section 9.5.1 "Space Object Automatic State Change Detection Algorithm Development," page 135).

9.3.4.7 Space Intelligence Preparation of the Battlespace

SWAT has a database and user interface to allow the space analyst to implement a full space IPB.

9.3.4.8 Satellite Characteristics Database

The SWAT software has a database of space object characteristics that is a combination of the Satellite Assessment Center (SatAC) Satellite Information Database (SID) [3], and additional satellite data researched by Paul Szymanski. This SWAT database

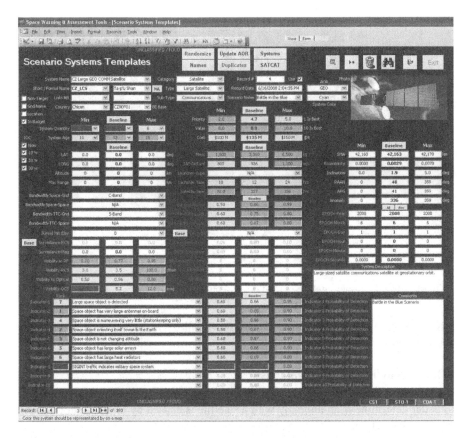

FIGURE 9.9   SWAT ACE game example system template.

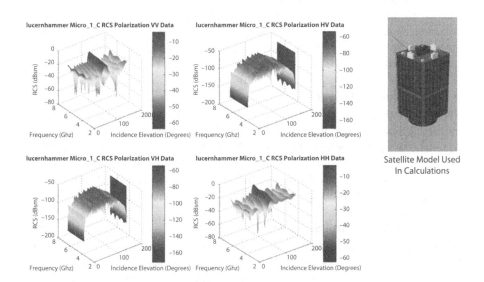

FIGURE 9.10   Example SWAT satellite RADAR cross-section.

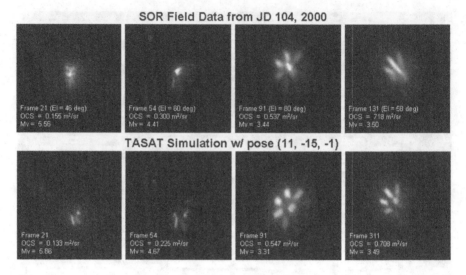

FIGURE 9.11   Example satellite optical cross-section data.

has extensive statistics generation capabilities to allow the user to gain insight into selected categories of space objects.

9.3.4.9   Satellite Failures Database

The SWAT software has a database of satellite failure history that enables the analyst to better determine if an anomaly with a satellite is caused by intentional, unintentional or natural forces.

9.3.4.10   Space INTEL Tasking Prioritization

The SWAT software has a space IPB tool that allows the analyst to task intelligence collection assets based on prioritized need. The INTEL tasking prioritization can be automatically generated by SWAT based on most probable adversary attack strategy that SWAT has estimated.

9.3.4.11   Blue COA Generator

SWAT software has extensive lists of possible space strategies, military objectives, vulnerable space centers of gravity (Blue and Red), and space principles of war. In addition, the Space Highest Information Value Assessment (SHIVA) [4] software tool invented by the author, automatically ranks satellites and terrestrial space-related elements for their value in assisting the flow of worldwide critical military data supporting the battlefield.

9.3.4.12   Space Control Scenario

A space control scenario has been developed based on exercise data that show the relationships of terrestrial battlefield actions to space events. This includes Operational Objectives, Tactical Objectives, Tactical Tasks, Success Criteria, and Success Indicators. Battlefield tempo for both space and terrestrial forces is contained

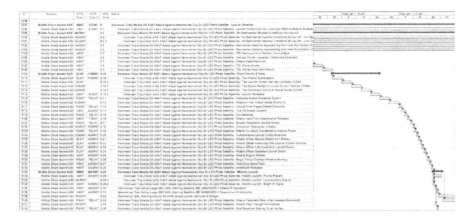

FIGURE 9.12   Example notional space scenario (partial).

in Microsoft Project scheduling software, along with extensive lists of intelligence indicators (see Section 9.7 in "Part 3 – Example Real-World Use of SWAT/SAW," page 149, also Figures 9.12 and 9.13).

9.3.4.13   NIIRS Space Equivalents

A NIIRS (National Imagery Interpretability Rating Scale) for terrestrial systems was extended for space systems. This gives the user a definition for quality of imagery data for both terrestrial and space systems. GRD means Ground Resolved Distance, and is used in a similar way for space systems. This rating system is similar to the Air Force Research Labs (AFRL) SORS (Space Object Rating Scale), except this Space NIIRS definition includes Ratings 1 and 2, which SORS does not (Table 9.5).

```
1.2.5.1.5.21    Chicanean Yuan Hsi Mobile Direct Ascent ASAT Manufacturing Centers Increased Wear On Roads at Sites
1.2.5.1.5.22    Chicanean Yuan Hsi Mobile Direct Ascent ASAT Manufacturing Centers Improved / New Roads at Sites
1.2.5.1.5.23    Chicanean Yuan Hsi Mobile Direct Ascent ASAT Manufacturing Centers Improved / New Parking at Sites
1.2.5.1.5.24    Chicanean Yuan Hsi Mobile Direct Ascent ASAT Manufacturing Centers Improved / New Railroad Tracks at Sites
1.2.5.1.5.25    Chicanean Yuan Hsi Mobile Direct Ascent ASAT Manufacturing Centers Improved / New Railroad Sidings at Sites
1.2.5.1.5.26    Chicanean Yuan Hsi Mobile Direct Ascent ASAT Manufacturing Centers Increased Disturbed Vegetation / Soil at Sites
1.2.5.1.5.27    Chicanean Yuan Hsi Mobile Direct Ascent ASAT Manufacturing Centers Different Communications Patterns To / From Sites
1.2.5.1.5.27.1  Chicanean Yuan Hsi Mobile Direct Ascent ASAT Manufacturing Centers Increased Communications Traffic To / From Sites
1.2.5.1.5.27.2  Chicanean Yuan Hsi Mobile Direct Ascent ASAT Manufacturing Centers Decreased (More Attempts to Hide) Communications Traffic To / From Sites
1.2.5.1.5.27.3  Chicanean Yuan Hsi Mobile Direct Ascent ASAT Manufacturing Centers No Net Increase or Decrease of Communications Traffic To / From Sites, But Changed Patterns
1.2.5.1.5.27.4  Chicanean Yuan Hsi Mobile Direct Ascent ASAT Manufacturing Centers Increased Encrypted Communications Traffic To / From Sites
1.2.5.1.5.28    Chicanean Yuan Hsi Mobile Direct Ascent ASAT Manufacturing Centers Increased Hours / New Shifts for Personnel at Sites
1.2.5.1.5.29    Chicanean Yuan Hsi Mobile Direct Ascent ASAT Manufacturing Centers Increased Number of Scientists & Engineers at Sites
1.2.5.1.5.30    Chicanean Yuan Hsi Mobile Direct Ascent ASAT Manufacturing Centers Increased Number of Military Personnel at Sites
1.2.5.1.5.31    Chicanean Yuan Hsi Mobile Direct Ascent ASAT Manufacturing Centers Increased Number of Military Personnel of Higher Ranks at Sites
1.2.5.1.5.32    Chicanean Yuan Hsi Mobile Direct Ascent ASAT Manufacturing Centers Increased Number of Foreign Personnel at Sites
1.2.5.1.5.33    Chicanean Yuan Hsi Mobile Direct Ascent ASAT Manufacturing Centers Increased Number of VIPs at Sites
1.2.5.1.5.34    Chicanean Yuan Hsi Mobile Direct Ascent ASAT Manufacturing Centers Increased Housing Demand In Local Area
1.2.5.1.5.35    Chicanean Yuan Hsi Mobile Direct Ascent ASAT Manufacturing Centers New / Expanded / Improved Housing Built On-Site
1.2.5.1.5.36    Chicanean Yuan Hsi Mobile Direct Ascent ASAT Manufacturing Centers New / Expanded / Improved Recreational Facilities On-Site
1.2.5.1.5.37    Chicanean Yuan Hsi Mobile Direct Ascent ASAT Manufacturing Centers Increased Food Intake
1.2.5.1.5.38    Chicanean Yuan Hsi Mobile Direct Ascent ASAT Manufacturing Centers Increased Power Consumption
1.2.5.1.5.39    Chicanean Yuan Hsi Mobile Direct Ascent ASAT Manufacturing Centers Increased Water Consumption
1.2.5.1.5.40    Chicanean Yuan Hsi Mobile Direct Ascent ASAT Manufacturing Centers Increased Sewer Outake
1.2.5.1.5.41    Chicanean Yuan Hsi Mobile Direct Ascent ASAT Manufacturing Centers Increased Refuse Outake
1.2.5.1.5.42    Chicanean Yuan Hsi Mobile Direct Ascent ASAT Manufacturing Centers Increased Smoke Plumes from Sites
1.2.5.1.5.43    Chicanean Yuan Hsi Mobile Direct Ascent ASAT Manufacturing Centers Increased Chemical Contamination at Sites
1.2.5.1.5.44    Chicanean Yuan Hsi Mobile Direct Ascent ASAT Manufacturing Centers New or Increased Settling / Effluents Ponds at Sites
1.2.5.1.5.45    Chicanean Yuan Hsi Mobile Direct Ascent ASAT Manufacturing Centers Increased Use of Data Processing Assets at Site
1.2.5.1.5.46    Chicanean Yuan Hsi Mobile Direct Ascent ASAT Manufacturing Centers Increased / Different Patterns of Thermal Images
1.2.5.1.5.47    Chicanean Yuan Hsi Mobile Direct Ascent ASAT Manufacturing Centers Increased Number of Large Mobile Vehicles with Erection Gantries at Sites
1.2.5.1.5.48    Chicanean Yuan Hsi Mobile Direct Ascent ASAT Manufacturing Centers Increased Number of Mobile Vehicles with Cooling at Sites
1.2.5.1.5.49    Chicanean Yuan Hsi Mobile Direct Ascent ASAT Manufacturing Centers Increased Number of Chemical Support Equipment at Sites
1.2.5.1.5.50    Chicanean Yuan Hsi Mobile Direct Ascent ASAT Manufacturing Centers Increased Number of Optical Test Equipment at Sites
1.2.5.1.5.51    Chicanean Yuan Hsi Mobile Direct Ascent ASAT Manufacturing Centers Increased Number of RF Test Equipment at Sites
1.2.5.1.5.52    Chicanean Yuan Hsi Mobile Direct Ascent ASAT Manufacturing Centers Increased Number of Electrical Test Equipment at Sites
1.2.5.1.5.53    Chicanean Yuan Hsi Mobile Direct Ascent ASAT Manufacturing Centers Increased Number of Optical Test Stands at Sites
1.2.5.1.5.54    Chicanean Yuan Hsi Mobile Direct Ascent ASAT Manufacturing Centers Increased Number of RF Test Stands at Sites
1.2.5.1.5.55    Chicanean Yuan Hsi Mobile Direct Ascent ASAT Manufacturing Centers Increased Number of Large Mobile Vehicle Storage Sheds at Sites
1.2.5.1.5.56    Chicanean Yuan Hsi Mobile Direct Ascent ASAT Manufacturing Centers Increased Security at Sites
```

FIGURE 9.13   Example notional space systems INTEL indicators (partial).

9.3.4.14   Space Objectives

Many space objectives were collected by the author over the last four decades from military exercises. In addition, many others were developed from translating terrestrial objectives to space-related ones. There are currently approximately 5,000 objectives in this SWAT database (Table 9.7).

9.3.4.15   Space Centers of Gravity

Table 9.6 gives a partial list of some of the more unique Centers Of Gravity (COG) associated with space systems and space control. These were derived from considerable strategic thought based on extensive space control analyses over 48 years and the unique characteristics of the space environment. It should be noted that a COG can be a physical location (a mountain pass or crowded satellite orbit) or a weakness in strategic doctrine concerning the use of space systems, etc. AFDD 2-2.1, defines a COG as, "Those characteristics, capabilities or sources of power from which a military force derives its freedom of action, physical strength or will to fight" [5]. Joint Publication 1-02 defines a COG as a "primary source of moral (i.e., political leadership, social dynamics, cultural values, or religion) or physical (i.e., military, industrial, or economic) strength from which a nation, alliance, or military force in a given strategic, operational, or tactical context derives its freedom of action, physical strength, or will to fight" [6].

9.4   ANALYSIS METHODOLOGY

9.4.1   SWAT State Change Algorithm

The SWAT software algorithm (Figure 9.14) that automatically classifies the mission of any unknown space object or determines state changes, essentially takes all of the available information on an unknown satellite (such as orbital elements (altitude, inclination, eccentricity, etc.), size, shape, stabilization (spinning, 3-axis stabilized), RCS, optical visual magnitude, mass, power, GEO orbital position, Delta-v, spin rate, etc.) (see Table 9.2 for a complete list of space object characteristics used in SWAT algorithms) and compares it to the same types of characteristics for satellites of known missions. SWAT calculates the "distance" in multi-dimensional space (each dimension is the value of a satellite characteristic) of an unknown space object to those of known mission. The unknown space object is assessed to be of the mission that is "closest" in this multi-dimensional space. Figure 9.14 illustrates this process with one satellite characteristic of inclination (of course, this would make a 1-dimensional chart—the chart is shown as two dimensions to make it easier for the viewer to understand the process). Blue boxes illustrate the range of values of inclination for weather satellites, while green circles show the range of inclination values for science satellites.

As can be seen in Figure 9.14, some of these ranges of inclination values merge for different missions. However, the preponderance of range values do show a statistical separation. The unknown space object inclination value is shown by the red asterisk. As can be readily viewed in this figure, the red asterisk is "closer" to the weather satellites inclination ranges than science satellites. Thus, SWAT assesses the unknown

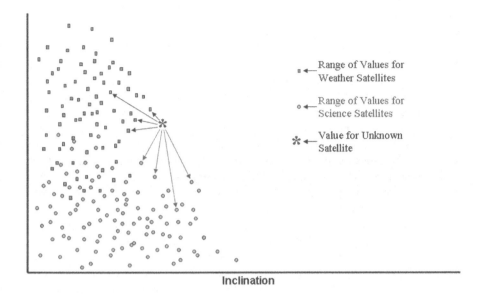

Inclination

FIGURE 9.14   SWAT state change algorithm.

space object as having the highest probability of being a weather satellite—at least as far as satellite inclination goes. SWAT then compares the other unknown space object characteristics to those of all possible space missions, normalizes and averages all of the results, and gives an overall score as to its most probable space mission. The user is also able to weight (bias) some satellite characteristics higher than others in the algorithm calculations for particular missions. A list of all possible missions and state changes, sorted by the most probable or highest mission or change first is then presented to the user to help him direct intelligence collection resources that will verify those most probable missions and state changes (SWAT automatically increases the intelligence collection priority ranking for those space objects needing verification as part of its Space Intelligence Preparation of the Battlespace (SIPB) algorithms).

9.5   EXAMPLE ANALYSES

9.5.1   SPACE OBJECT AUTOMATIC STATE CHANGE DETECTION ALGORITHM DEVELOPMENT

This algorithm in SWAT automatically detects changes of state for all space objects in the USSPACECOM satellite catalog. This entire process includes three phases of analysis. Phase 1 concerns space objects whose designated mission does not correspond to the characteristics of other space objects of that same mission. This might indicate a war-reserve or secondary mission for the space object that has not previously been detected. This first phase analyzes all *current* space object characteristics data. Phase 2 uses the same algorithms as the first, but compares all space object missions to all *current* and *past* characteristics data. This second phase of the SWAT state change algorithms involves tracking the changes of individual space objects from their respective historical range of values. Because of the large size of the SWAT characteristics historical database (currently 4,117,708 records), SWAT allows the user to set the minimum date a space object was launched (currently set at the year 1999) before it can be analyzed (Phase 1 algorithms use all space object data for the currently analyzed period). This keeps the calculations run times in SWAT at a lower level when importing new space object data.

SWAT Phase 3 state change algorithms analyze the history of individual space object characteristics for any significant deviation from norm, without regard to their stated mission. SWAT automatically detects these differences as new space object data are received. The space object characteristics data that are tracked are RCS, optical cross-section (visual magnitude), orbital elements, stabilization type (spinning or 3-axis), object shape (sphere, cylinder, box), length, width, height, mass, and spin rate (see Table 9.2 for a complete list of space object characteristics used in SWAT algorithms). These characteristics are imported into SWAT from the SID databases and archives. This importing is performed every two weeks, which started with July, 2007 data to the present date. During the importation process, SWAT algorithms correlate all objects to their respective missions and themselves, and then presents the user with a prioritized list of those objects that are least correlated (Figure 9.5).

Various new coding and filtering functions have been added as a result of extensive testing with real space object data from the SatAC SID databases and JSpOC data. Certain user-selectable satellite missions (e.g., Interplanetary, Test-Craft, and Science, etc.), orbital status (e.g. LUNAR IMPACT), satellite status (e.g., Engineering C/O), and select countries can be excluded from the state change calculations (Figure 9.15). Also,

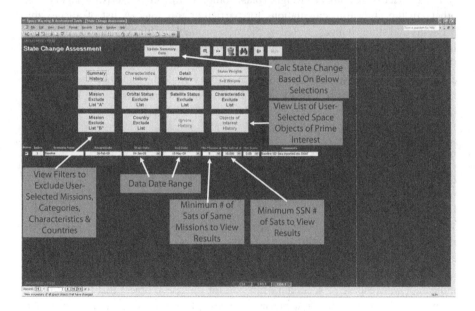

FIGURE 9.15 SWAT auto state change user filter switchboard.

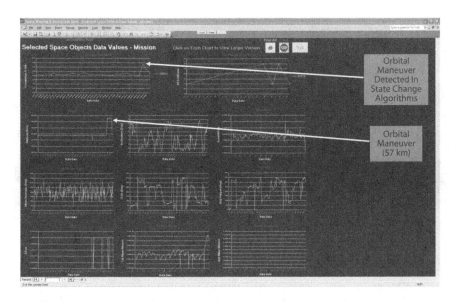

FIGURE 9.16   SWAT auto State Change Example—ECHOSTAR 5.

extensive database cleanup is required before processing, such as mission definition inconsistencies from JSpOC (e.g. "Science" vs. "Scientific" missions). In addition, new user interfaces were developed with additional summarized data to assist
the user in determining why a particular space object has been identified by SWAT
as the most changed of all objects in the space catalog (Figures 9.16–Figure 9.18).

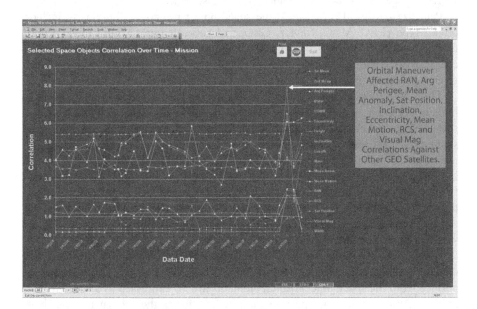

FIGURE 9.17   SWAT auto state change example correlations over time graph—ECHOSTAR 5.

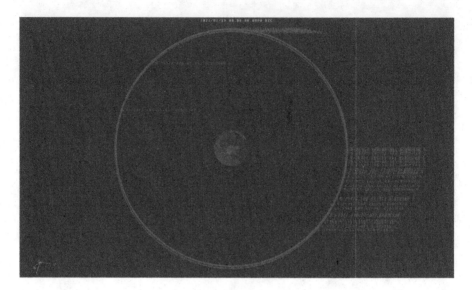

FIGURE 9.18 SWAT auto state change example orbital history graphic—ECHOSTAR 5.

Several interesting space objects with significant differences from other objects of the same purported mission have been identified during this SWAT testing and have been reported to government sponsors. In addition, new coding was inserted into SWAT to duplicate the state change, mission assessment, and COA assessment functions with notional data to allow SWAT demonstrations at unclassified levels which segregates the real space object data from exercise data.

Paul Szymanski also expanded the space object mission definitions to include orbital type (LEO, MEO, MOLY, GEO, and sub-categories within these overall orbital types: see Table 9.8) to enable distinguishing significant changed object characteristics within a mission while excluding pre-defined orbital characteristics grouping differences.

The SWAT state change algorithms have to account for some subtle differences in data. For example, the previous SPARKS database listed the spin rate for those space objects that are not 3-axis stabilized. All other objects had blank fields for their spin rates. These blank fields have now been filled with zeroes to reflect that they do not have a spin at all to them, instead of a blank field symbolizing that no spin rate calculations would be performed for that space object in the state change algorithms. In contrast, some SID records for space object mass are listed as 0. This would tell the SWAT algorithm that those categories of satellites actually have zero mass. All of these zero values have to be detected and changed to null so those calculations would not bias the state change algorithms.

In addition, the databases have been improved for some subtle differences in data processing. For example, data such as RAN, mean anomaly and argument of perigee from the space object orbital elements are all expressed as a value from 0 to 359 degrees. This is also true for satellite position in the geosynchronous belt. In the original SWAT state change algorithms, a change from 359 to 1 degrees

TABLE 9.8
Space Defense Regions Definitions

Region	Region Definition
SDR GEO	Space Defense Region Geosynchronous
SDR GEO ASIA	Space Defense Region Geosynchronous over Asia
SDR GEO EU	Space Defense Region Geosynchronous over Europe
SDR GEO ME	Space Defense Region Geosynchronous over the Middle East
SDR GEO US	Space Defense Region Geosynchronous over the United States
SDR GEO-G-A	Space Defense Region Graveyard Orbit Above Geosynchronous
SDR GEO-G-B	Space Defense Region Graveyard Orbit Below Geosynchronous
SDR GEO-I	Space Defense Region Geosynchronous Inclined
SDR HEO	Space Defense Region Above Geosynchronous (High Earth Orbit)
SDR LEO-E	Space Defense Region Low Earth Orbit Highly Eccentric
SDR LEO-H	Space Defense Region Low Earth Orbit - High (>600 and <5,000 km)
SDR LEO-L	Space Defense Region Low Earth Orbit - Low (<=500 km)
SDR LEO-M	Space Defense Region Low Earth Orbit - Medium (>500 and <=600 km)
SDR LEO-R	Space Defense Region Low Earth Orbit Retrograde
SDR LEO-S	Space Defense Region Low Earth Orbit Sun-Synchronous
SDR MEO	Space Defense Region Medium Earth Orbit (>=5,000 and <25,000 km)
SDR MOLY	Space Defense Region Molniya
SDR NOE	Space Defense Region No Orbital Elements

would be marked as a large change (358 degrees). In reality, this would actually only be a small change of 2 degrees if it occurred over this arbitrary 359-0 degrees boundary. New SWAT coding takes these odd cases and calculates the correct change in degrees, assuring that no change is more than 180 degrees. In addition, duplicated data were eliminated from the statistical calculations to remove bias when there is no new updates (some space objects may not have had an updated element set or RCS value in the two weeks between SWAT data analysis runs).

Example SWAT state change outputs are in Tables 9.9 and 9.10. These two tables show top changed space objects where the changes were mostly influenced by optical sensor data.

9.5.2   OPTICAL DATA IMPACT ON AUTOMATIC STATE CHANGE DETECTION ALGORITHM

9.5.2.1   Optical Data Introduction

Mr. Szymanski imported government-supplied and archival space object optical characteristics data into SWAT. In addition, space object "flashing" data were imported from internet databases of worldwide user measurements. This "flashing" data give the period of rotation of those space objects visible to these amateur observers. Its main advantage is the extensive data starting in 1958 to the present that includes LEO, MEO, and GEO space objects. We currently have approximately

TABLE 9.9
Example Optical State Changes

SATCAT No	SatName	Mission	Country	Orbit	Comments
17083	GORIZONT 13	COMM-CIVIL	Russia	SDR GEO	Along with GORIZONT 7, GORIZONT 13 is the dimmest GORIZONT in the sky (13) vs. visual magnitude of 6 for other GORIZONT's.
16667	COSMOS 1738	COMM-CIVIL	Russia	SDR GEO	Visual magnitude slightly dimmer (13.2) than many other satellites of its class (5.5–12.5).
16650	BRAZILSAT 2	COMM-CIVIL	Brazil	SDR GEO	Visual magnitude much brighter (1 - flash) than other satellites of its class (4–14).
23267	COSMOS 2291	COMM-MIL	Russia	SDR GEO	Visual magnitude much brighter (6) than other satellites of its class—US (11–11.6).
20523	INTELSAT 603	COMM-CIVIL	INTELSAT	SDR GEO	Along with 21653 (INTELSAT 605) Visual Magnitude much brighter (3) than other satellites of its class (6 - 14.7). At the time, the Intelsat 6 series were the largest commercial spacecraft ever built.
15946	RADUGA 16	COMM-CIVIL	Russia	SDR GEO	Visual magnitude slightly dimmer (13.8) than many other satellites of its class (5.5–13.2).
26069	COSMOS 2369	ELINT	Russia	SDR LEO	Along with 28352 (another ELINT) visual magnitude slightly brighter (4.5) than most other satellites of its class (5.5–5.6).
15398	COSMOS 1610	NAVSAT	Russia	SDR LEO	Visual magnitude slightly brighter (4) than other satellites of its class (5–10).
22971	SL-14 R/B	SL-14 R/B	Russia	SDR LEO	Visual magnitude much dimmer (9.8) than other satellites of its class (5–6.5).
11165	COSMOS 1066	METSAT	Russia	SDR LEO	Visual magnitude slightly dimmer (6.7) than other satellites of its class (5.3–6.4).
15774	SL-12 R/B(AUX MOTOR)	Rocket Body	Russia	SDR LEO-H	Flash period (5.7) significantly less than other SL-12 R/B(AUX MOTOR) SSN: 15338 (15).
25415	ORBCOMM FM 19	COMM-MOBIL	ORBCOMM	SDR LEO-H	Visual magnitude much dimmer (9) than other satellites of its class (Iridium—6.5; GLOBALSTAR—5.5).
25116	ORBCOMM FM 9	COMM-MOBIL	ORBCOMM	SDR LEO-H	Visual magnitude much dimmer (9) than other satellites of its class (Iridium—6.5; GLOBALSTAR—5.5).
16191	**METEOR 3-1**	**METSAT**	**Russia**	**SDR LEO-H**	**Visual magnitude much dimmer (6.7) than most other satellites of its class (5.3-5.8) [possibly because it is a new model; METEOR 3 vs. METEOR 1 or 2].**
15930	COSMOS 1670	RORSAT	Russia	SDR LEO-H	Visual magnitude slightly brighter (6) than other satellites of its class (5.6).
11084	**COSMOS 1045**	**OCEANOGRPY**	**Russia**	**SDR LEO-H**	**Visual magnitude slightly dimmer (6) than five other satellites of its class (5.5); note object is extremely stable in its orbit.**
11671	COSMOS 1151	ELINT	Russia	SDR LEO-L	Visual magnitude slightly dimmer (5.5) than most other satellites of its class (5.2–5.4).
25396	TMSAT	EARTH-RES	Thailand	SDR LEO-S	Visual magnitude much dimmer (9) than other satellites of its class (4.5–6.9).
17199	ARIANE 1 DEB	ARIANE 1 DEB	France	SDR LEO-S	Flash period much higher than other ARIANE 1 DEB.
27430	**HAIYANG 1**	**METSAT**	**China**	**SDR LEO-S**	**Flash period more than doubles on 6/16/2008.**
21935	**SL-12 DEB**	**SL-12 DEB**	**Russia**	**SDR MEO**	**Radical change in flash period.**
13080	COSMOS 1341	MSL-WARN	Russia	SDR MEO	Visual magnitude slightly dimmer (5.5) than most other satellites of its class (1–5). Flash period much lower (3.4) than others of its class (7–47).
21855	COSMOS 2179 (GLONASS)	NAVSAT	Russia	SDR MEO	Visual magnitude very much dimmer (10.9) than other satellites of its class (1.5–3).

TABLE 9.10
Example Optical State Changes (Continued)

SATCAT No	SatName	Mission	Country	Orbit	Comments
17083	GORIZONT 13	COMM-CIVIL	Russia	SDR GEO	Along with GORIZONT 7, GORIZONT 13 is the dimmest GORIZONT in the sky (13) vs. visual magnitude of 6 for other GORIZONTs.
16667	COSMOS 1738	COMM-CIVIL	Russia	SDR GEO	Visual magnitude slightly dimmer (13.2) than many other satellites of its class (5.5–12.5).
16650	BRAZILSAT 2	COMM-CIVIL	Brazil	SDR GEO	Visual magnitude much brighter (1–flash) than other satellites of its class (4–14).
23267	COSMOS 2291	COMM-MIL	Russia	SDR GEO	Visual magnitude much brighter (6) than other satellites of its class—US (11–11.6).
20523	INTELSAT 603	COMM-CIVIL	INTELSAT	SDR GEO	Along with 21653 (INTELSAT 605), visual magnitude much brighter (3) than other satellites of its class (6–14.7). At the time, the Intelsat 6 series were the largest commercial spacecrafts ever built.
15946	RADUGA 16	COMM-CIVIL	Russia	SDR GEO	Visual magnitude slightly dimmer (13.8) than many other satellites of its class (5.5–13.2).
26069	COSMOS 2369	ELINT	Russia	SDR LEO	Along with 28352 (another ELINT), visual magnitude slightly brighter (4.5) than most other satellites of its class (5–5.6).
15398	COSMOS 1610	NAVSAT	Russia	SDR LEO	Visual magnitude slightly brighter (4) than other satellites of its class (5–10).
22971	SL–14 R/B	SL–14 R/B	Russia	SDR LEO	Visual magnitude much dimmer (9.8) than other satellites of its class (5–6.5).
11165	COSMOS 1066	METSAT	Russia	SDR LEO	Visual magnitude slightly dimmer (6.7) than other satellites of its class (5.3–6.4).

138,000 optical measurements in the SWAT databases. The SOR (Starfire Optical Range) databases that are across 80 Excel spreadsheets were also formatted and imported, along with Russian optical data obtained from Maui. The SOR Color Photometry GEO Catalog data runs from 1996 to 2007 (39,936 Records), has 92 GEO-only satellites and 13 optical filter measurements. In addition, Maui provided Russian data measured between 1987–2001 (33,095 records) with 22 GEO-only satellites with three optical filters. The Belgian Astronomical Association [7] provides optical flashing data (along with some visual magnitude data) on its website from 1958–2008 (65,284 Records) for 2,064 space objects (all orbits).

SWAT state change algorithms (see Section 9.5.1) were run both with and without optical data used to characterize the space objects. In previous analyses, RCS data had a significant positive impact on the SWAT automatic mission assessment algorithms (see Section 9.3.4.1). It was hoped that optical cross-section (visual magnitude) might provide a similar impact. The analysis date ranges ran from Jan–Oct 2008 (October was the latest date we had for JSpOC RCS data at the time). SWAT algorithms used the latest available data for each two-week analysis step date.

9.5.2.2   Optical Data Results

Optical data had a significant impact on state change rankings using the SWAT algorithms. After applying the full optical data, 33 percent of space object state change scores increased (improved detection), 50 percent of space object change scores decreased (addition of optical data helped stabilized erratic data), and 17 percent of space object change scores remained unchanged. Higher scores indicates higher state change. It appears that, for the limited set of optical data obtained (mostly old GEO data for visual magnitudes), there was still a significant benefit to including all of the optical data. In addition, individual space object differences can be derived directly from the optical data (Tables 9.9 and 9.10). For example, in Table 9.9, the optical signature for object 16191 (METEOR 3-1) indicates a possible difference in visual magnitude between METEOR 1 and 2 satellites versus the later METEOR 3 version. In addition, objects 27430 (HAIYANG 1) and 21935 (SL-12 launch vehicle debris) both had radical changes in flash period during the analysis timespan. This might indicate a major change in tumbling direction for these space objects, or in time of war, it might indicate a hidden change in mission to a war reserve mode. All of these mission differences and state changes were automatically detected by the SWAT state change algorithms and immediately displayed to the user [8].

Part 2: Example Analyses

9.6   SATELLITE ATTACK WARNING DISPLAYS

New user displays were developed in SWAT to delineate the current space situation in terms of predictive battlespace awareness. These AVIS (Altitude Versus Inclination Survey [9]) plots attempt to simplify the space situation view for the warfighter by only illustrating orbital and other changes for space objects, while "fixing" the actual orbital movements to prevent user data overload (Figures 9.7 through 9.19). In Figure 9.7 each dot represents an individual space object. These space objects are distributed on the graph only according to their altitude and

FIGURE 9.19   Example SWAT SAW space map symbols. See Appendix for a complete list of SWAT-developed space icons.

inclination—the two most important factors when assessing the amount of Delta-v it would take for any of the objects to maneuver closer to its neighbor. Even if the space object is on the other side of the Earth from a target satellite, in the orbital space defined by Figure 9.7, it is considered "close" because the attacker can choose the time and phasing of its attack (using bi-elliptical transfer, if necessary). This chart assumes an RPO (Rendezvous and Proximity Operations) type of attack, where the attacker essentially matches the orbit of the target (co-orbital attack). For the case of a glancing attack where the orbits do not match, but the orbital tracks meet (COSMOS 2251 and Iridium 33 collision which had a 12-degree difference in inclination, but the same orbital altitude [10]), this AVIS chart would show all objects of the same altitude as a horizontal line that have the possibility of collision/attack. This glancing attack mode is the subject of subsequent SAW display development. Figure 9.20 shows the same data as Figure 9.7, only with the more traditional view of satellites around the Earth. As can be readily observed, the AVIS charts concept simplifies considerably the user data overload and the ability to detect an adversary positioning for a future attack.

In Figure 9.7 the color of the labels denotes the country affiliation of the satellite, rocket booster, or space debris. One can see that certain altitude-inclination regions have a preponderance of one country's space objects over another. When a potential threat country's object is in a usually associated Blue zone, this can cause an analyst to further explore these threat objects. Also, Figure 9.21 illustrates notional Delta-v "rings" around certain satellites that will be the subject of future AVIS chart developments.

Figure 9.22 shows the Figure 9.7 display after the user has zoomed in on a region of interest. Figure 9.18 shows some of the space-related military icons that were developed in accordance with Mil-Std-2525B [11] to display space objects according to country affiliation, mission, and operational status. 220 new space icons were drawn and added to the 44 already existing in Mil-Std-2525B (Figure 9.19; see Appendix for a complete list of SWAT-developed space icons). In addition, the trailing lines associated with each space object in this display denote their relative positions over a

FIGURE 9.20   Traditional orbital view with same data as previous SWAT choke point map.

time period set by the user. In a sense, some of them appear to be "falling out of the sky" in this display as they lose altitude. Figure 9.23 shows an additional zooming in of Figure 9.7, where additional amplifying information is attached to each icon giving name, mission, remaining life, estimated Delta-v remaining, state change, etc. This additional information is also based on Mil-Std-2525B, as extended by the author to space objects. As part of this process, space defense regions were defined for all orbits of interest to characterize localized space threats (Table 9.9). Figure 9.24 shows

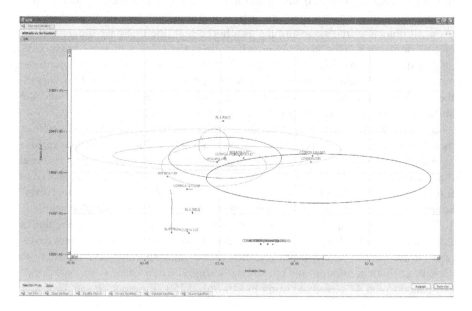

FIGURE 9.21   SWAT choke point map—notional maneuver envelopes.

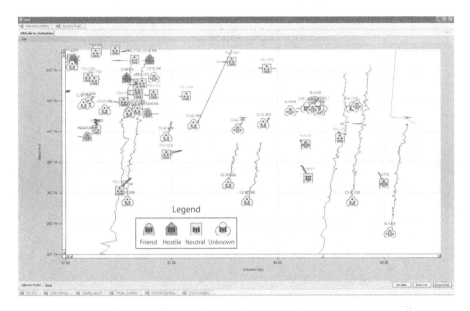

FIGURE 9.22   SWAT choke point map—example icons.

the same AVIS chart with the Space Defense Identifications Zone (SDIZ) designations turned on. These SDIZ definitions are extensions of air defense terminology
and help in managing space control assets and warning tripwires.

Figures 9.25–9.27 illustrate how a simulated attack against GPS navigation satellites would look in the AVIS displays. In the center of Figure 9.25, red launch vehicle
rocket boosters appear to be maneuvering towards GPS satellites. The boosters'

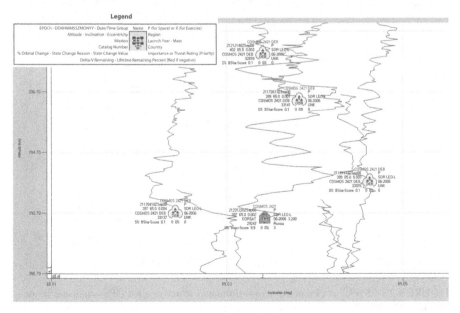

FIGURE 9.23   SWAT choke point map—space objects with characteristics data.

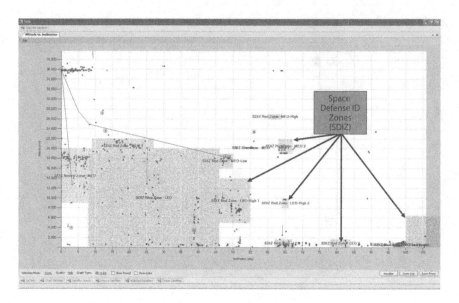

FIGURE 9.24 SWAT choke point map—Space Defense Identification Zones (SDIZ).

previous history have always been at altitudes and inclinations different than GPS, thus defining a Blue zone where US navigation satellites operate. Figure 9.26 zooms in on the previous figure, and shows the red boosters' change in orbital values by the trace lines leading towards GPS satellites. Figure 9.27 zooms in even further than the previous figure in the AVIS display to show the final approach of the red ASATs

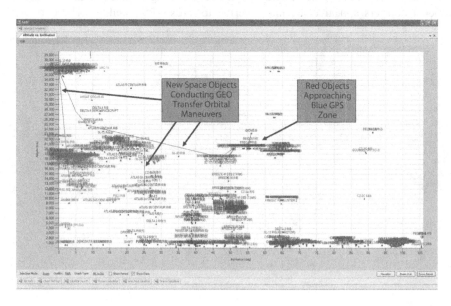

FIGURE 9.25 SWAT choke point map—simulated attack against GPS—unusual maneuvers towards "Blue" zone.

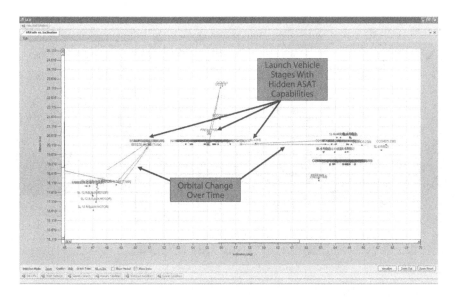

FIGURE 9.26   SWAT choke point map—simulated attack against GPS—war reserve mode for "dead" stages.

and their Mil-Std-2525D derived symbology with associated text values expressing technical characteristics and SWAT-calculated state change scores for each space object. For the red rocket stages in this notional scenario, the state change scores are large because the RCS have radically changed to reflect changes in attitude of these stages as they orient themselves for orbital maneuvers.

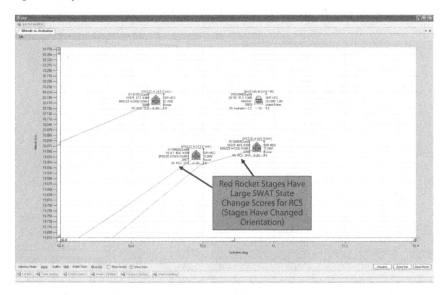

FIGURE 9.27   SWAT choke point map—simulated attack against GPS—multiple attacks against one GPS target.

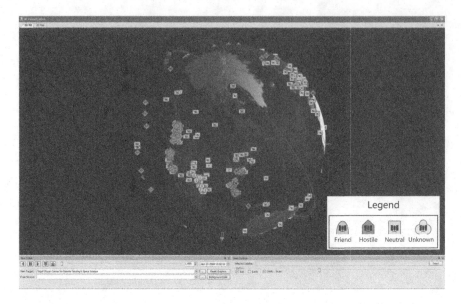

FIGURE 9.28 SWAT choke point map—3D visualization of selected space objects and terrestrial sites (ACE wargame).

A flat map view along with 3-dimensional globe views were developed to enable terrestrial space surveillance and attack envelopes to be assessed (Figures 9.28 and 9.29). Mr. Szymanski also participated in the Advanced Concept Event (ACE) Space Wargame test conducted 8–11 December 2008, as an effort to test SWAT concepts

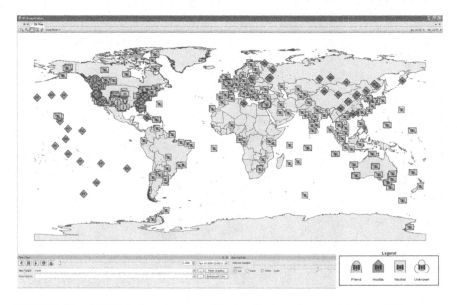

FIGURE 9.29 SWAT choke point map—2D visualization of terrestrial sites (ACE wargame + worldwide space sites).

and user interfaces. This was a success for the SWAT algorithms and led to many new ideas for SWAT improvements.

Part 3: Example Real-World Use of SWAT/SAW

9.7 AUTOMATIC SPACE SYSTEMS SCENARIO GENERATION TOOL

In order to test SWAT algorithms, coding, databases, and user interfaces, a very robust space scenario is required. In addition, the Advanced Concept Event (ACE) exercise where SWAT was tested and demonstrated required a scenario. Because of this, the SWAT scenario construction capabilities (based on the JWID 2000 Exercise [12]) were started early in the development cycle. For flexibility in generating multiple scenarios for broad SWAT testing, this coding automatically constructs and outputs scenarios from SWAT. The database outputs for this space scenario were offered to exercise participants for use in developing their individual program events for ACE. Figure 9.30 shows an example map for the ACE scenario that is in SWAT. Figure 9.31 gives one of the Areas Of Responsibility (AORs) graphics for the geosynchronous belt with the satellite color coding designated

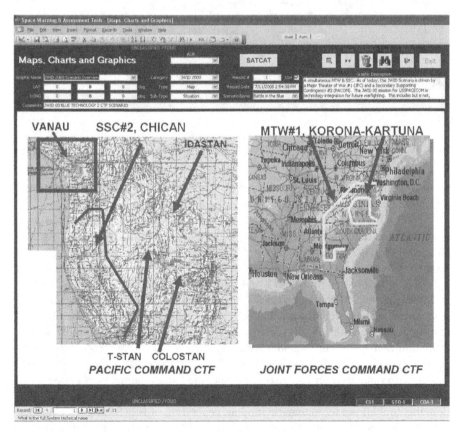

FIGURE 9.30   SWAT ACE game example maps.

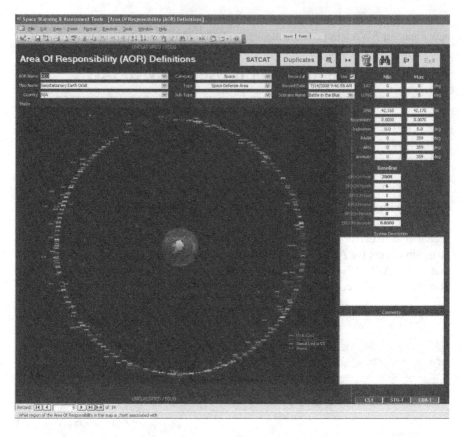

FIGURE 9.31   SWAT ACE game example AORs for space.

according to possible threat countries. New concepts for GEO space defense areas and space defense regions were also developed (Table 9.8 and Figure 9.32).

Figure 9.9 shows one (out of a total of 393) of the generic space systems templates that are randomized between user-selectable limits and then used to generate a total of 2,282 systems for use in ACE. Each template has a generic picture of a space system associated with it (out of a total of 969 generic pictures that were obtained and formatted for SWAT), and a generic proper name based on a database in SWAT of 169,903 random names derived from a worldwide geographic database. Every space system has associated intelligence indicators and probabilities of detection. Each country in this scenario tool is give a random space budget (within minimum and maximum bounds) which is used to "buy" the space systems by priority order until all the money is spent. This way each side in the wargame will end up with random (user can also select specific values) numbers of systems and system types so that the value of differing force mixes can be evaluated. In addition, this tool allows for automatic random generation of BE (Basic Encyclopedia) numbers for data tracking purposes. Figures 9.33 and 9.34 give a partial list of military objectives and space events for ACE that were entered into a Microsoft

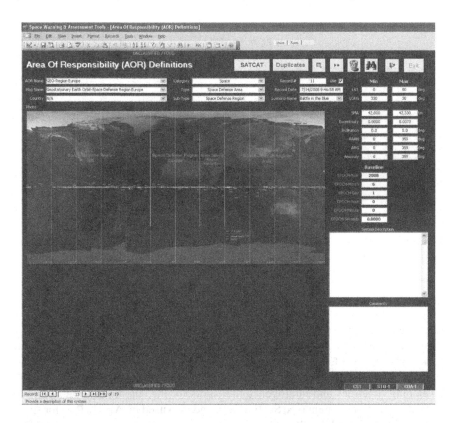

FIGURE 9.32   SWAT ACE game geosynchronous space defense regions definition.

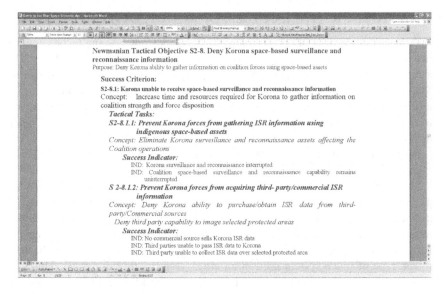

FIGURE 9.33   SWAT ACE scenario military objectives example (partial).

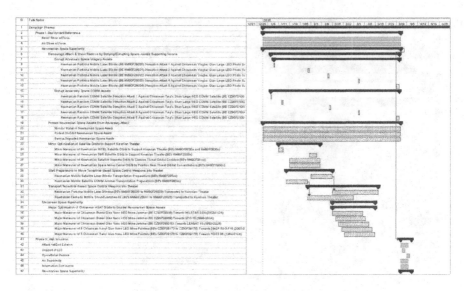

FIGURE 9.34   SWAT ACE scenario battle tempo (partial).

Project© database. Such a scenario timeline should have more detailed orbital simulation time data inserted at a future date, that can then be placed in Monte Carlo software to determine a statistical range of time possibilities. This scenario tool was also used to generate generic space weapon systems that were used in developing timelines and intelligence indicators for the automatic COA detection tool software development (see Section 9.8), and also used by AFRL/RIE for space command and control development.

9.8   AUTOMATIC SPACE COURSE OF ACTION DETECTION TOOL

SWAT has extensive lists of potential adversary attack strategies and tactics against space systems. For each of these potential attack strategies, SWAT gives the intelligence indicators that would be present if the strategy was being implemented against US and allied space systems. The SWAT user enters any INTEL indicators that are being currently observed, no matter how insignificant, and SWAT automatically matches these with the most probable attack strategy or ASAT type that an adversary may be implementing towards us. SWAT can also automatically increase the INTEL tasking priority for those indicators that would best verify which strategy is being implemented by a potential adversary.

Table 9.11 is a small sample of some of the intelligence indicators that SWAT uses in determining whether our space systems are under attack, and what kind of attack it is. These, and other intelligence indicators, are linked in SWAT to potential adversary attack strategies, tactics, or COAs. Mr. Szymanski developed an extensive (over 6,262 events) Microsoft Project space scenario with detailed INTEL indicators for each type of space system used in an attack (Figures 9.12 and 9.13). This was imported into SWAT to enable an INTEL indicator checklist for differing types of attack. This space scenario indicators work also supports INTEL assessment efforts at AFRL/RIE.

TABLE 9.11
Example Space Strategies INTEL Indicators (Partial)

Activation of sensor protective covers

Activation of terrestrial war reserve facilities related to space capabilities

Advantageous solar phase angle

Appendages moving

Approach trajectories outside normally employed orbits

Approach trajectories outside the field of regard of the target's on-board sensors

Approach trajectories when the Sun/Moon/Earth is in the background of a target's sensors

Are a small number of Blue and Gray satellites experiencing anomalies over a long time period

Are a small number of Blue and Gray satellites losing contact with terrestrial controllers

Are a small number of new Red satellites appearing in orbit

Are a small number of Red satellites changing orientation

Are a small number of Red satellites changing shape

Are a small number of Red satellites changing thermal signatures

Are a small number of Red satellites concentrating towards potential Blue and Gray satellites

Are Red ASAT forces appearing to line up in a sequence of timed attacks against Blue and Gray assets

Are Red forces capable of attacking space-related terrestrial sites in Blue countries appearing to line up in a sequence of timed attacks

Are Red SIGINT assets appearing to line up in a sequence of timed operations against Blue and Gray Communications assets

Are space surveillance sensors currently tracking the satellite

Are terrestrial centers currently commanding the satellite

Are there a small number of new satellite launches from Red facilities

Are there indications of Red aircraft activities that appear to concentrate on space-related terrestrial sites around the world

Attacks against space surveillance assets

Attitude changes

Attitude changes to turn sensitive components away from perceived threats

Bent structure

Communications or telemetry frequencies that can be jammed or spoofed

Counter-strikes against satellite launch facilities

Counter-strikes against space control weapons

Counter-strikes against terrestrial non-space-related assets

Counter-strikes against terrestrial satellite terminals

Current battery charge state

Dead satellites acting alive

Effluent expulsions

Electrical distribution status

Electrostatic discharge phenomena

Escape of sub-satellites from main satellite buss

Escort satellite activation in vicinity of defended satellite

Expulsion of chaff or aerosols for protection

Fluids being transferred

GEO satellites changing orbital position

Has the satellite maneuvered

Haze on optics

(*Continued*)

TABLE 9.11 *(Continued)*
Example Space Strategies INTEL Indicators (Partial)

Heaters warming thruster fuel

Holes in structure

Increased charging of satellite batteries

Increased security at terrestrial facilities related to space capabilities

Increased space surveillance activities

Increased use of gray space systems

Is satellite currently rebooting its controllers

Is satellite in a safe hold condition

Is the satellite currently maneuvering

Is the satellite preparing to maneuver

Is the weapon system preparing/powering up for use

Just after loss of contact with adversary satellite ground controllers

Just after loss of contact with adversary space surveillance assets

Lens covers moving

Is the adversary making prominent demonstrations of attack in one sector of space, while quietly
attacking in another, less observed sector of space

Maneuver changes

Material degradation

Missing parts from spacecraft

Near a satellite's high power antennas

Near a satellite's thrusters

New satellite launches

New space objects appearing in orbit

Outgassing

Overheated electric power converters

Pass over potential Red ASAT assets

Pass over Red space surveillance assets

Periods of solar eclipse for satellites

Periods when a satellite has a low battery charge

RF signal traffic level

Ruptured battery cells

Satellite constellation changes

Satellite Current Antenna Pointing Direction

Satellite Current On-Board Processor State

Satellite Current Orientation Attitude

Satellite Current RF Signal Characteristics

Satellite Current Sensor State

Satellite Current Solar Panel Pointing Direction

Satellite Orbital Position Over Time

Seal failures

Solar array motors status

Solar flares & meteor showers

Solar panel age

(Continued)

TABLE 9.11 *(Continued)*
Example Space Strategies INTEL Indicators (Partial)

Solar panel broken interconnect wires
Solar panel fogging of covering materials
Solar panel holes
Solar panel orientation
Solar panel power output
Solar panel soldering problems
Solar panels being stowed
Space-to-Space links in use
Sub-satellites exiting main satellite
Sun / Moon blind spots for on-board sensors
Theater supported alert status
Thruster plumes
Times of cloud cover/weather/natural disasters for terrestrial-based space surveillance systems
Times of cloud cover/weather/natural disasters for terrestrial-based space weapons systems
Times when the satellite passes through space radiation belts
Venting of gasses from parts of the structure
Weapons warming up
Where is main axis pointed towards
Where is sensor suite pointed towards

Figure 9.4 is a screenshot from the SWAT software user input form that automatically assesses intelligence indicators to determine what is the most probable attack strategy, tactic, and associated conflict level that an adversary may be implementing against US and allied space systems. The bottom of the screen gives the intelligence indicators that the user has selected from a pull-down list that reflects the current status of space systems. The top portion of the screen lists the most likely adversary attack strategies, tactics (COAs), and conflict levels that are currently being implemented, in order of most probable first. There is a separate briefing that defines each of these strategies, tactics, and conflict levels. An extensive list (2,506) of intelligence indicators keywords and phrases was also developed to assist the user in locating the appropriate indicator.

Part 4: Final Summary

9.9 CONCLUSIONS

Space object change detection in SWAT appears to be able to automatically determine when a space object, whether live or dead, has changed in some way (orbital or characteristics), and presents to the user a prioritized list for further space surveillance/ SOI assessment (for both Red and Blue space objects). It also appears that space object optical data can have a significant impact on this state change algorithm. The optical data appear to allow distinguishing satellite types and significant characteristics differences. The SWAT choke point/SAW maps have generally been well-received by the space community as a unique way to assess the space situation. They deserve further development. The intelligence indicators data developed for the SWAT Red COA detection algorithms have been well-received by the community and extensively used by AFRL/RIE contractors and proved critical for program success. The actual ability to detect Red COAs can only be fully proven in an actual space war. However, space war simulations, wargames, and exercises can help determine its validity in the future.

9.10   RECOMMENDATIONS/NEXT STEPS

9.10.1   Operationalize Space Object Change Detection Algorithms

SWAT space object change detection appears to be working and providing significant data on a bi-weekly basis. This data should be passed to JSpOC operators to help them in prioritizing their space surveillance/SOI requirements.

9.10.2   Obtain Space Object Optical Sensor Data

Space object optical sensor data had a significant impact on the space object change detection algorithms in SWAT. However, this data were mostly on geosynchronous satellites, and the latest data measurement date was in 2007. More recent and extensive data are required. For example, Los Alamos National Labs has ground-based optical data, and somewhere there exists previous SBV space-based sensors that may provide MEO and LEO data. The STSS program can provide additional, current data. Finally, the GEODSS system may have optical data measurements that can be obtained.

9.10.3   Evolve Space Choke Point Maps

The SWAT choke point/SAW maps have already proven useful during ACE exercise testing. However, this was just the beginning, and additional displays need to be designed. For example, the AVIS display shows when space objects have similar altitudes and inclinations, making them "close" in orbital space in terms of the amount of Delta-v required to rendezvous with a potential target. In the case of the COSMOS 2251 and Iridium 33 collision [13], which had a 12-degree difference in inclination, but the same orbital altitude, this would be displayed on a common horizontal line in AVIS. Obviously, a better way must be developed to illustrate "glancing" attacks against satellites where the inclination does not have to be the same for both objects. Also, it is intended to design Delta-v vs. transit time maps in some way on these charts.

9.10.4   Develop Space Wargaming Toolsets

The best way to test the SWAT choke point/SAW maps and the Red COA detection algorithms is in theoretical space wargames. SWAT already has automatic space scenario generation tools, so these need to be expanded to include a full wargame testing capability. This would include tools to help develop strategies, and optimize Delta-v vs. time during synchronized, multiple attacks employing weapon systems of varied phenomenologies. This would test possible space attack envelopes, sequences, and tempos while training future warfighters in detecting surprise attacks in space.

Ultimately, this provides Red and Blue optimal attack strategies based on SID (SatAC Satellite Information Database) satellite characteristics and sensor orbital data.

Part 5: The Other "SWAT"

9.11   ARMY SWAT

The Army Space and Missile Defense Center has developed a different software product called Space Wargaming Analysis Tool. This was developed by the United States Army years after the Air Force SWAT tool was first developed and deployed, and apparently accidently shares the same acronym. A description of this army tool is given below, which is based upon the US Army Space and Missile Defense Center Public Affairs Fact Sheet [14].

9.11.1   Army Space Wargaming Analysis Tool (SWAT)

Space Wargaming Analysis Tool is an analytical tool that enables the rapid and dynamic execution of wargaming courses of action and generates critical data that can be used to inform commanders or decision makers regarding space concepts, capabilities, CONOPs and TTPs in environments with and without space-based capabilities.

When warfighters go into battle, they make and revise strategies on a continual basis. They often protect national security, which is why they need to make sure they are prepared to defend against whatever the enemy brings their way. Warfighters need a tool that supports the development of multi-domain scenarios to provide situational understanding of the potential impacts of strategic and tactical decisions. Enter the Space Wargaming Analysis Tool, a rapidly deployable, quick-scenario generation-and-execution tool that enables high-level analysis of single or multiple platforms supporting the armed forces in a wargaming environment. The Space Wargaming Analysis Tool enables high- level analysis of single or multiple platforms supporting space, air, and ground maneuver (red and blue) forces in a wargaming environment, while providing data reduction in real-time allowing the user to gain an understanding of the impacts of scripted and dynamic assets against the enemy.

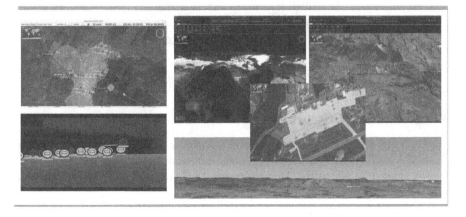

- Dynamically inject multiple platforms
- Create systems in a scenario or master scenario event list (also supports the dynamic injection)
- Worldwind elevation data
- Distributed interactive simulation protocol distributed systems
- Built-in post processing that provides the warfighter with detailed data during the study
- Graphical user interface that utilizes NASA Worldwind map engine data to create a user interface that helps the warfighter get a better laydown of the area of interest
- Developing augmented reality view for users

US ARMY SPACE AND MISSILE DEFENSE COMMAND

9.11.2 SIMULATING THE BATTLEFIELD

The US Army Space and Missile Defense Command is developing and deploying the Space Wargaming Analysis Tool to gain real-time understanding of the effects of its actions against an adversary. This government-off-the-shelf tool will be available to all Department of Defense organizations wanting to utilize a space wargaming environment.

The Space Wargaming Analysis Tool's primary function is to generate Red team versus Blue team wargames, where two sides battle each other virtually with scripted or dynamically injected interactions and the effects of those actions play out against each other. The tool is unique in that it enables the dynamic deployment of battle assets—divisions, brigades, battalions, companies, platoons and fire teams—in real time or faster than real time, providing further insight into potential scenarios. These scenarios can include uncommon events, such as degradation or jamming of communications, which force users to adjust strategies on the fly.

SWAT imports satellite flight paths and satellite constellations so that all domains, including space, are included in the wargaming analysis. A comprehensive understanding of all outcomes requires a comprehensive input of scenarios, and Space Wargaming Analysis Tool achieves just that.

9.11.3 Multiplatform Deployment

The program is accessible on a laptop; however, the team is working to condense its operability to a tablet to make it even easier to use.

SWAT users get a bird's eye view on their laptops and in augmented reality. By using augmented reality headsets and peripherals, they see a virtual sand table that enables easy collaboration and communication.

The user interfaces in both augmented reality and desktop views are based on NASA's WorldWind map engine data. This is an open-source, virtual globe that allows developers to quickly create interactive and accurate visualizations of the earth. The user interface helps the warfighter understand their area of interest.

9.11.4 Future Developments

In addition to enhancing mobility, the team is incorporating inclement weather, reconnaissance, navigation, and other variables and scenarios that soldiers may encounter.

The modern warfighter needs as much information as possible before engaging with adversarial forces. The Space Wargaming Analysis Tool quickly provides planners with that awareness. The tool helps ensure troops are ready to complete their critical missions whenever they are on the field of battle.

Highlights
- Develop "base case" scenarios consisting of space, air, and ground entities with scripted paths
- Prototype ground entity options include division, brigade, battalion, company, platoon, fire teams, and individual entities (e.g., tanks)

- Prototype space entity options include satellites and satellite constellations
- Currently utilizes Satellite Orbit Analysis Program file data for satellite constellation paths
- Scenarios include planned events that are scripted into the base case scenario. Events would include (but not be limited to) degradation of communication, jamming, spoofing, and damage
- WorldWind 3D Terrain elevation engine
- Up to three active frequencies per entity
- Red versus Blue
- Interactive mode that allows for dynamic injection of entities in real time
- Space Wargaming Analysis Tool supports a faster-than- real-time function and pause at the specified break point
- Detailed events describing space, air, and ground force states at different points of the scenario
- Supports augmented reality view of battle

The Space Wargaming Analysis Tool team is researching new capabilities to add to the program, not only for its current customer, but for future customers as well.

Distribution A 1221-01

For more information, please contact:
USASMDC Public Affairs Office
P.O. Box 1500
Huntsville, AL 35807
Phone: 256-955-3887
www.linkedin.com/company/armysmdc
www.instagram.com/armysmdc
www.smdc.army.mil
www.facebook.com/armysmdc
www.twitter.com/armysmdc
www.flickr.com/armysmdc
www.youtube.com/armysmdc

Space Glossary List

Glossary	Definition	Source
Space Defense	All defensive measures designed to destroy attacking enemy vehicles (including missiles) while in space, or to nullify or reduce the effectiveness of such attack. (1) All defensive measures designed to destroy or nullify attacking enemy spacecraft and missiles and also negate hostile space systems. (2) An inclusive term encompassing space defense and ballistic missile defense. See also aerospace defense (also see Joint Pub 1-02).	Joint Pub 3-01.1
Space Forces	The space and terrestrial systems, equipment, facilities, organizations, and personnel necessary to access, use, and, if directed, control space for national security (JP 1-02).	DODD 3100.10, July 9, 1999; DODD 5101.2, June 3, 2003; AFDD 2-2.1 2, August 2004
Space Power	The total strength of a nation's capabilities to conduct and influence activities to, in, through, and from the space medium to achieve its objectives.	DODD 3100.10, July 9, 1999; DODD 5101.2, June 3, 2003
Space Transportation	The capability for responsive, reliable, flexible, and assured access to space in peace, crisis, and through appropriate levels of conflict commensurate with national security requirements.	
Space Control	Combat and combat support operations to ensure freedom of action in space for the United States and its allies and, when directed, deny an adversary freedom of action in space. The space control mission area includes surveillance of space; protection of US and friendly space systems; prevention of an adversary's ability to use space systems and services for purposes hostile to US national security interests; negation of space systems and services used for purposes hostile to US national security interests; and directly supporting battle management, command, control, communications, and intelligence.	DODD 3100.10, July 9, 1999
Active Space Defense	Direct defensive action taken to destroy, nullify, or reduce the effectiveness of hostile space actions. It includes the use of anti-satellite weapon systems, defensive counter space weapons, electronic warfare, and other available weapons not primarily used in a space defense role. See also Space Defense.	Modified from Joint Pub 3-01.1
Space Control Operations	The employment of space forces, supported by air, ground and naval forces, as appropriate, to achieve military objectives in vital areas of concern to space systems. Such operations include destruction of enemy in-space assets, space-related ground systems and surface-to-space forces (launch), interdiction of enemy space operations, protection of vital space lines of communication (links from ground to space to ground), and the establishment of local military superiority in areas of space operations.	Modified from Joint Pub 3-01.1

(Continued)

Space Glossary List *(Continued)*

Glossary	Definition	Source
Space Defense Action Area	An orbit and the space around it within which friendly spacecraft or surface-to-space weapons are normally given precedence in operations except under specified conditions. Also see Space Defense Operations Area.	Modified from Joint Pub 3-01.1
Space Defense Area	(1) A specifically defined orbit for which space defense must be planned and provided. (2) An orbit and a region surrounding it of defined dimensions designated by the appropriate agency within which the ready control of space-borne vehicles is required in the interest of national security during an space defense emergency.	Modified from Joint Pub 3-01.1
Space Defense Artillery	Weapons and equipment for actively combating space targets from the ground.	Modified from Joint Pub 3-01.1
Space Defense Battle Zone	A volume of space surrounding a space defense fire unit or defended area, extending to a specified orbital altitude and inclination, in which the fire unit commander will engage and destroy targets not identified as friendly under criteria established by higher headquarters. In other words, this would be a free-fire zone around a defended satellite.	Modified from Joint Pub 3-01.1
Space Defense Control Center	The principal information, communications, and operations center from which all spacecraft, anti-satellite operations, space defense artillery, guided missiles, and space warning functions of a specific area of space defense responsibility are supervised and coordinated. Also called Space Defense Operations Center.	Modified from Joint Pub 3-01.1
Space Defense Division	A geographic subdivision of a Space Defense Region. Also see Space Defense Sector.	Modified from Joint Pub 3-01.1
Space Defense Emergency	An emergency condition, declared by the Commander in Chief, USSTRATCOM, that exists when attack upon space systems of interest to the United States by hostile spacecraft, missiles, or ground weapons, is considered probable, is imminent, or is taking place.	Modified from Joint Pub 3-01.1
Space Defense Identification Zone	Orbital space of defined parameters within which the ready identification, location, and control of spaceborne vehicles is required. Also called SDIZ. Also see Space Defense Operations Area.	Modified from Joint Pub 3-01.1
Space Defense Operations Area	An area and the orbital space around it within which procedures are established to minimize mutual interference between space defense and other operations; it may include designation of one or more of the following: Space Defense Action Area, Space Defense Area; Space Defense Identification Zone, and/or firepower umbrella.	Modified from Joint Pub 3-01.1
Space Defense Region	An orbital subdivision of a Space Defense Area.	Modified from Joint Pub 3-01.1
Space Defense Sector	An orbital subdivision of a Space Defense Region. Also see Space Defense Division.	Modified from Joint Pub 3-01.1

(Continued)

Space Glossary List *(Continued)*

Glossary	Definition	Source
Space Sovereignty	A nation's inherent right to exercise absolute control and authority over the orbital space near its satellites. Also see Space Sovereignty Mission.	Modified from Joint Pub 3-01.1
Space Sovereignty Mission	The integrated tasks of surveillance and control, the execution of which enforces a nation's authority over the orbital space near its satellites. Also see Space Sovereignty.	Modified from Joint Pub 3-01.1
Space Deconfliction In The Combat Zone	A process used to increase combat effectiveness by promoting the safe, efficient, and flexible use of space systems. Space deconfliction is provided in order to prevent fratricide, enhance space defense operations, and permit greater flexibility of operations. Space deconfliction does not infringe on the authority vested in commanders to approve, disapprove, or deny combat operations. Also called Combat Space Deconfliction; Space Deconfliction.	Modified from Joint Pub 3-01.1
Space Control Sector	A sub-element of the space control area, established to facilitate the control of the overall orbit. Space control sector boundaries normally coincide with space defense organization subdivision boundaries. Space control sectors are designated in accordance with procedures and guidance contained in the space control plan in consideration of service component and allied space control capabilities and requirements.	Modified from Joint Pub 3-01.1
Space Autonomous Operation	In space defense, the mode of operation assumed by a space system after it has lost all communications with human controllers. The space system assumes full responsibility for control of weapons and engagement of hostile targets, based in accordance with on-board surveillance and weapon system control logic. This automatic state may occur on a regular basis due to orbital movements outside regions of ground coverage and control.	Modified from Joint Pub 3-01.1
Broadcast-Controlled Space Interception	An interception in which the interceptor is given a continuous broadcast of information concerning the space defense situation and effects interception without further control.	Modified from Joint Pub 3-01.1
Space Centralized Control	In space defense, the control mode whereby a higher echelon makes direct target assignments to fire units. See also Decentralized Control.	Modified from Joint Pub 3-01.1
Close-Controlled Space Interception	An interception in which the interceptor is continuously controlled to a position from which the target is within local sensor range.	Modified from Joint Pub 3-01.1
Space Decentralized Control	In space defense, the normal mode whereby a higher echelon monitors unit actions, making direct target assignments to units only when necessary to ensure proper fire distribution or to prevent engagement of friendly spacecraft. See also Centralized Control.	Modified from Joint Pub 3-01.1

Space Glossary List *(Continued)*

Glossary	Definition	Source
Passive Space Defense	All measures, other than active space defense, taken to reduce the probability of, and to minimize the effects of, damage to space systems caused by hostile action without the intention of taking the initiative. These measures include camouflage, deception, dispersion, and the use of protective construction and design. See also Space Defense.	Modified from Joint Pub 3-01.1
Space Point Defense	The defense or protection of special vital elements, orbital positions (geosynchronous slots, and advantageous orbits, such as sun-synchronous) and installations (e.g., command and control facilities, space launch facilities, tracking, telemetry and control facilities, space surveillance sensors, and high-value satellites).	Modified from Joint Pub 3-01.1
Space Positive Control	A method of space control which relies on positive identification, tracking, and situation assessment of spacecraft within a space defense area, conducted with electronic means by an agency having the authority and responsibility therein.	Modified from Joint Pub 3-01.1
Space Weapon Engagement Zone	In space defense, orbital space of defined altitude and inclination within which the responsibility for engagement of space threats normally rests with a particular weapon system. Also called SWEZ. (1) Direct-Ascent Engagement Zone (DAEZ): In space defense, that orbital space of defined altitude and inclination within which the responsibility for engagement of space threats normally rests with a direct-ascent anti-satellite system of terrestrial launch origin. (2) Directed Energy Engagement Zone (DEEZ): In space defense, that orbital space of defined altitude and inclination within which the responsibility for engagement of space threats normally rests with a directed energy (laser or microwave) ASAT or electronic warfare system of terrestrial location. (3) Electronic Warfare Engagement Zone (EWEZ): In space defense, that orbital space of defined altitude and inclination within which the responsibility for engagement of space threats normally rests with an electronic warfare system of terrestrial location. (4) Close Attack Engagement Zone (CAEZ): In space defense, that orbital space of defined altitude and inclination within which the responsibility for engagement of space threats normally rests with an ASAT system that is stationed within 10 kilometers of its target. (5) Long Range Engagement Zone (LREZ): In space defense, that orbital space of defined altitude and inclination within which the responsibility for engagement of space threats normally rests with long-range space defense weapons that are space-based but are normally stationed at more than 10 kilometers from its target. (6) Joint Engagement Zone (JEZ): In space defense, that orbital space of defined altitude and inclination within which multiple space defense systems (from both terrestrial and space-based locations) are simultaneously employed to engage space targets.	Modified from Joint Pub 3-01.1

(Continued)

Space Glossary List *(Continued)*

Glossary	Definition	Source
Space Weapons Assignment	In space defense, the process by which weapons are assigned to individual space weapons controllers for use in accomplishing an assigned mission.	Modified from Joint Pub 3-01.1
Space Weapons Free	In space defense, a weapon control order imposing a status whereby weapons systems may be fired at any target in orbital space of defined altitude and inclination, not positively recognized as friendly. See also Weapons Hold; Weapons Tight.	Modified from Joint Pub 3-01.1
Space Weapons Hold	In space defense, a weapon control order imposing a status whereby weapons systems may only be fired in self-defense or in response to a formal order. See also Weapons Free; Weapons Tight.	Modified from Joint Pub 3-01.1
Space Weapons Tight	In space defense, a weapon control order imposing a status whereby weapons systems may be fired only at targets recognized as hostile. See also Weapons Free; Weapons Hold.	Modified from Joint Pub 3-01.1
Counterspace	Those offensive and defensive operations conducted by air, land, sea, space, special operations, and information forces with the objective of gaining and maintaining control of activities conducted in or through the space environment. (AFDD 2-2)	AFDD 2-2.1 2, August 2004
Defensive Counterspace	Operations to preserve US/friendly ability to exploit space to its advantage via active and passive actions to protect friendly space-related capabilities from adversary attack or interference. Also called DCS.	AFDD 2-2.1 2, August 2004
Offensive Counterspace	Operations to preclude an adversary from exploiting space to their advantage. Also called OCS. (AFDD 2-2.1)	AFDD 2-2.1 2, August 2004
Space	A medium like the land, sea, and air within which military activities shall be conducted to achieve US national security objectives. (JP 1-02)	AFDD 2-2.1 2, August 2004
Space Capability	(1) The ability of a space asset to accomplish a mission. (2) The ability of a terrestrial-based asset to accomplish a mission in space (e.g., a ground-based or airborne laser capable of negating a satellite). (JP 1-02) [The ability of a space asset or system to accomplish a mission.]	AFDD 2-2.1 2, August 2004
Space Coordinating Authority	The single authority designated, by the supported combatant commander or JFC, to coordinate joint theater space operations and to integrate space capabilities thereby facilitating unity of the theater/JOA space effort. Also called SCA. (AFDD 2-2.1)	AFDD 2-2.1 2, August 2004

(Continued)

Space Glossary List *(Continued)*

Glossary	Definition	Source
Space Control	Combat, combat support, and combat service support operations to ensure freedom of action in space for the US and its allies and, when directed, deny an adversary freedom of action in space. The space control mission area includes surveillance of space; protection of US and friendly space systems; prevention of an adversary's ability to use space systems and services for purposes hostile to US national security interests; negation of space systems and services used for purposes hostile to US national security interests; and directly supporting battle management, command, control, communications, and intelligence. (JP 1-02) [Operations to assure the friendly use of the space environment while denying its use to the enemy. Achieved through offensive and defensive counterspace carried out to gain and maintain control of activities conducted in or through the space environment.] (AFDD 2-2).	AFDD 2-2.1 2, August 2004
Space Environment	The region beginning at the lower boundary of the earth's ionosphere (approximately 50 km) and extending outward that contains solid particles (asteroids and meteoroids), energetic charged particles (ions, protons, electrons, etc.), and electromagnetic and ionizing radiation (X-rays, extreme ultraviolet, gamma rays, etc.). (JP 1-02)	AFDD 2-2.1 2, August 2004
Space Situation Awareness	The knowledge and intelligence that provides the planner, commander, and executor with sufficient awareness of objects, activities, and the environment to enable COA development. This involves characterizing, as completely as possible, the space capabilities operating within the terrestrial and space environments. Space situation awareness forms the foundation for all space activities, and is the enabler for counterspace operations. Also called SSA. (AFDD 2-2.1)	AFDD 2-2.1 2, August 2004
Space Superiority	The degree of dominance in space of one force over another that permits the conduct of operations by the former and its related land, sea, air, space, and special operations forces at a given time and place without prohibitive interference by the opposing force. (JP 1-02) [The degree of control necessary to employ, maneuver, and engage space forces while denying the same capability to an adversary.]	DODD 3100.10, July 9, 1999; AFDD 2-2.1 2, August 2004
Space Support	Combat service support operations to deploy and sustain military and intelligence systems in space. The space support mission area includes launching and deploying space vehicles, maintaining and sustaining spacecraft on-orbit, and deorbiting and recovering space vehicles, if required. (JP 1-02) [Those operations conducted with the objective of deploying, sustaining, and augmenting elements or capabilities of military space systems. Space support consists of spacelift and on-orbit support.] [AFDD 2-2]	DODD 3100.10, July 9, 1999; AFDD 2-2.1 2, August 2004

(Continued)

Space Glossary List *(Continued)*

Glossary	Definition	Source
Space Support Team	A team of space operations experts provided by the commander, US Space Command (or one of the space component commands and augmented by national agencies, as required) upon request of a geographic combatant commander to assist the supported commander in integrating space power into the terrestrial campaign. Also called SST. (JP 1-02)	AFDD 2-2.1 2, August 2004
Space Systems	All of the devices and organizations forming the space network. These consist of: spacecraft; mission packages(s); ground stations; data links among spacecraft, mission or user terminals, which may include initial reception, processing, and exploitation; launch systems; and directly related supporting infrastructure, including space surveillance and battle management and/or command, control, communications and computers. (JP 1-02)	DODD 3100.10, July 9, 1999; DODD 5101.2, June 3, 2003; AFDD 2-2.1 2, August 2004
Suppression of Adversary Counterspace Capabilities	Suppression that neutralizes or negates an adversary offensive counterspace system through deception, denial, disruption, degradation, and/or destruction. These operations can target ground, air, missile, or space threats in response to an attack or threat of attack. Also called SACC. (AFDD 2-2.1)	AFDD 2-2.1 2, August 2004
Space Element	A platform in which astrodynamics is the primary principle governing its movement through its environment.	AFDD 2-2, 23 August 1998
Space Power	The capability to exploit civil, commercial, intelligence, and national security space systems and associated infrastructure to support national security strategy and national objectives from peacetime through combat operations.	AFDD 2-2, 23 August 1998
Space System	A system with a major functional component which operates in the space environment or which, by convention, is so designated. It usually includes a space element, a link element, and a terrestrial element.	AFDD 2-2, 23 August 1998
Counterspace	Those operations conducted with the objective of gaining and maintaining control of activities conducted in or through the space environment.	AFPAM 14-118 5 JUNE 2001
Military Space Forces	Those national, civil, and commercial space systems and associated infrastructure that establish space power and are employed by the military to achieve national security objectives. Space forces include space-based systems, ground-based systems for tracking and controlling objects in space and transiting through space, launch systems that deliver space elements, and people who operate, maintain, or support those systems. Terrestrial-based forces operate below 100 kilometers. Space-based forces operate above 100 kilometers.	AFPAM 14-118 5 JUNE 2001

(Continued)

Space Glossary List *(Continued)*

Glossary	Definition	Source
Space Control	Assures the friendly use of the space environment while denying its use to the enemy. Achieved through counterspace missions carried out to gain and maintain control of activities conducted in or through the space environment.	AFPAM 14-118 5 JUNE 2001
Space Force Application	Attacks against terrestrial-based targets carried out by military weapons systems operating in space. (AFDD 2-2)	AFPAM 14-118 5 JUNE 2001
Space Force Enhancement	Those operations conducted from space with the objective of enabling or supporting terrestrial-based forces. (AFDD 2-2)	AFPAM 14-118 5 JUNE 2001
Space Power	The capability to exploit space forces to support national security strategy and achieve national security objectives.	AFPAM 14-118 5 JUNE 2001
Space Superiority	Degree of control necessary to employ, maneuver, and engage space forces while denying the same capability to an adversary.	AFPAM 14-118 5 JUNE 2001
Space Support	Those operations conducted with the objective of sustaining, surging, and reconstituting elements or capabilities of military space systems. Space support consists of spacelift and on-orbit support.	AFPAM 14-118 5 JUNE 2001
Space Stations	Space stations are large orbiting structures designed to support manned operations for extended periods of time (months to years). Space stations have performed both military and civilian objectives.	JPL Mission and Spacecraft Library; October 2006

Acronym List [15]

Glossary	Definition
AADO	Assured Access Duty Officer
ACE	Advanced Concept Event (AFRL/RD military exercise)
ACTA	Applied Cognitive Task Analysis
AFTRS	Air Force Tactical Receive System
AGI	Analytical Graphics Inc.
AOC	Air Operations Center
ASAT	Anti-Satellite
ASW	Astrodynamic Safety Workstation
AVIS	Altitude Versus Inclination Survey
AWC	Advanced Warning Capability
BDA	Battle Damage Assessment
BMDS	Ballistic Missile Defense System

C2PC	Command And Control Personal Computer
CA	Conjunction Analysis
CCAS	Consolidated Capability Assessment Schedule
CCIR	Commander Critical Information Requirements
CCOMS	Correlation Center Output Message Set
CDM	Critical Decision Method
CFM	Cognitive Function Model
CMMA	Collection Management Mission Application
COA	Course Of Action
COD	Combat Ops Division
COLT	CCAS Online Tool
COP	Common Operating Picture
COTA	Cognitively Oriented Task Analysis
CP3	Core Process 3
CPS	Communication Processing System
CTA	Cognitive Task Analysis
DCO	Defense Connect Online
DEFSMAC	Defense Special Missile And Aerospace Center
DICE	Distributed C4ISR Environment
DIRSPACEFOR	Director, Space Forces
DNA	Decompose, Network And Assess
DS4	Director, Space Forces
ESC	Electronic Systems Command
FE	Force Enhancement Cell
FIDO	Flight Duty Officer
FLTC	Future Long Term Challenge
FRAGO	Fragmentary Order
FRPC	Fast Reaction Procedures Checklist
GALE	Generic Area Limitation Environment
GCCS-I3	Global Command and Control System Integrated Imagery And Intelligence
GCCS-J	Global Command and Control System—Joint
GDTA	Goal Directed Task Analysis
GEODSS	Ground-Based Electro-Optical Deep Space Surveillance
GIANT	GPS Interference And Navigation Tool
GMLRS	Guided Multiple Launch Rocket System
GOTS	Government Off-The-Shelf
GP	General Perturbation
GPSOC	Global Positioning System Operations Center
GSM	Global Summary Messages
GSSC	Global SATCOM Support Center
HTA	Hierarchical Task Analysis
ICS	Interacting Cognitive Subsystems
IO	Intelligence Officer
IPL	Image Product Library
IPOE	Intelligence Preparation of the Operational Environment

ISR	Intelligence Surveillance Reconnaissance
ISRD	Intelligence Surveillance Reconnaissance Division
ISSA	Integrated Space Situation Awareness
IWPC	Information Warfare Planning Capability
IWS	InfoWorkSpace
JCIO	JSpOC Capabilities Integration Office
JFCC	Joint Functional Component Command
JFCC SPACE	Joint Forces Combatant Commander Space
JFQ	Joint Forces Quarterly
JMSP	Joint Master Space Plan
JSARS	JSpOC Situation Awareness And Response System
JSIP	JFCC Space Integrated Prototype
JSOP	Joint Space Operations Plan
JTAGS	Joint Tactical Ground Station
JWIS	Joint Weather Impacts System
KADS	Knowledge Analysis And Documentation System
LISN	Launch Information Support Network
MCRS	Mission-Critical Reporting System
MCSB	Mission Control Station Backup
METOC-INT	METeorological and OCeanographic INTelligence
MILSTD	Military Standard
mIRC	mInternet Relay Chat
MLS	Multi-Level Security
MOPS	Measures Of Performance
MSP	Master Space Plan
MW	Missile Warning
MWDO	Missile Warning Duty Officer
NAVSOC	Naval Satellite Operations Center
NDO	Network Duty Officer
NIPRNET	Non-Secure Internet Protocol Router Network
NORTHCOM	Northern Command
NROC	National Reconnaissance Office Command Center
NSG	National System for Geospatial Intelligence
NSTR	Nothing Significant To Report
OCS	Optical Cross Section
ONIR	Overhead Non-Imaging Infrared
OODA	Observe Orient Decide Act
OPORD	Operational Order
OPSCAP	Operations Capable
OSA	Orbital Safety Analyst
OSC	Orbital Sciences Corporation
PARI	Precursor Action Result Interpretation
PBA	Predictive Battlespace Awareness
PDOP	Positional Dilution Of Precision
PDS-M	Processing and Display Subsystem Migration
PFPS	Portable Flight Planning Software

RCS	Radar Cross-Section
RAIDRS	Rapid Attack Identification and Reporting System
RD	Directed Energy Directorate
RH	Human Effectiveness Directorate
RHCV	Battlespace Visualization Branch
RI	Information Directorate
RPO	Rendezvous an Proximity Operations
RV	Space Vehicles Directorate
RY	AFRL Sensors Directorate
SADO	Space Situation Awareness Duty Officer
SADS	Situational Awareness Distribution Server
SAFIRE	Space Situation Awareness Fusion Intelligent Research Environment
SatAC	Satellite Assessment Center
SATRAN	Satellite-Reconnaissance-Advanced-Notice
SAW	Satellite Attack Warning
SBMCS	Space Battle Management Command Systems
SBV	Space Based Visible
SCA	Space Control Authority
SCIS	Survivable Communication Integration System
SCOPES	Space Common Operating Picture Exploitation System
SDDO	Space Defense Duty Officer
SERTS	Space Event and Re-Entry Tracking Software
SET	Space Event Technician
SHAC	Space High Accuracy Catalog
SHF	Super-High-Frequency
SID	Satellite Information Database
SIDC	Space Innovation And Development Center
SIGINT	Signals Intelligence
SIS	Space Intelligence Squadron
SOD	Space Operations Directive
SODO	Space Operations Duty Officer
SOI	Space Object Identification
SONET	Synchronous Optical Networking
SOPG	Space Operations Group
SOPS	Satellite Operating Squadrons
SOR	Starfire Optical Range
SP	Special Perturbations
SPARKS	Space Power Analysis and Requirements Keystone Software
SPINS	Special Instructions
SPSC	Space Control Center
SSA	Space Situation Awareness
SSAFE	Space Situational Awareness Foundational Enterprise
SSMF	Standard Survivable Message Format
SSR	Space Support Request
ST	Subscriber Terminal
STACS	Space Threat Analysis And Characterization System

STEED	Satellite Threat Evaluation Environment for Defensive Counterspace
STO	Space Tasking Order
STSS	Space Tracking and Surveillance System
SWAT	Space Warfare Analysis Tools
SWS	Space Warning Squadron
TACDAR	Tactical Detecting and Reporting
TDKC	The Design Knowledge Company
TIDO	Theater Intelligence Duty Officer
TIREM	Terrain Integrated Rough Earth Model
TKS	Task-Knowledge Structures
TOPO	Trajectory Operations Officer
TT&P	Tactics, Techniques and Procedures
UDOP	User-Defined Operating Picture
USNDS	United States Nuclear Detonation Detection System
USV	Unified Space Vault
WCSS	Work-Centered Support System

Source: Modified from "Cognitive Task Analysis for the Joint Space Operations Center (JSpOC)"; AFRL-RH-WP-TR-2009-00; The Design Knowledge Company; March 2009.

SWAT-Developed Space Icons [16]

Missile Launcher								
	MISSILE LAUNCHER-Friend.bmp	MISSILE LAUNCHER-Hostile.bmp	MISSILE LAUNCHER-Neutral.bmp	MISSILE LAUNCHER-Unknown.bmp				
Space Rocket Body								
	Missile Space Rocket Body-Friend.bmp	Missile Space Rocket Body-Hostile.bmp	Missile Space Rocket Body-Neutral.bmp	Missile Space Rocket Body-Unknown.bmp				
Space-Based ASAT								
	Satellite ASAT-Friend.bmp	Satellite ASAT-Hostile.bmp	Satellite ASAT-Neutral.bmp	Satellite ASAT-Unknown.bmp	Satellite ASAT-Dead-Friend.bmp	Satellite ASAT-Dead-Hostile.bmp	Satellite ASAT-Dead-Neutral.bmp	Satellite ASAT-Dead-Unknown.bmp
Space-Based ASAT (Directed Energy)								
	Satellite ASAT-DE-Friend.bmp	Satellite ASAT-DE-Hostile.bmp	Satellite ASAT-DE-Neutral.bmp	Satellite ASAT-DE-Unknown.bmp				
Space-Based ASAT (Electronic Warfare)								
	Satellite ASAT-EW-Friend.bmp	Satellite ASAT-EW-Hostile.bmp	Satellite ASAT-EW-Neutral.bmp	Satellite ASAT-EW-Unknown.bmp				

Space-Based ASAT (Kinetic Energy)								
	Satellite ASAT-KE-Friend.bmp	Satellite ASAT-KE-Hostile.bmp	Satellite ASAT-KE-Neutral.bmp	Satellite ASAT-KE-Unknown.bmp				
Space-Based ASAT (Space Mine) [RPO with target: kinetic, laser, jamming, paint, etc.]								
	Satellite ASAT-Mine-Friend.bmp	Satellite ASAT-Mine-Hostile.bmp	Satellite ASAT-Mine-Neutral.bmp	Satellite ASAT-Mine-Unknown.bmp				
ASAT Test Target								
	Satellite ASAT-Target-Friend.bmp	Satellite ASAT-Target-Hostile.bmp	Satellite ASAT-Target-Neutral.bmp	Satellite ASAT-Target-Unknown.bmp				
Satellite (Civil)								
	Satellite Civil-Friend.bmp	Satellite Civil-Hostile.bmp	Satellite Civil-Neutral.bmp	Satellite Civil-Unknown.bmp	Satellite Civil-Dead-Friend.bmp	Satellite Civil-Dead-Hostile.bmp	Satellite Civil-Dead-Neutral.bmp	Satellite Civil-Dead-Unknown.bmp

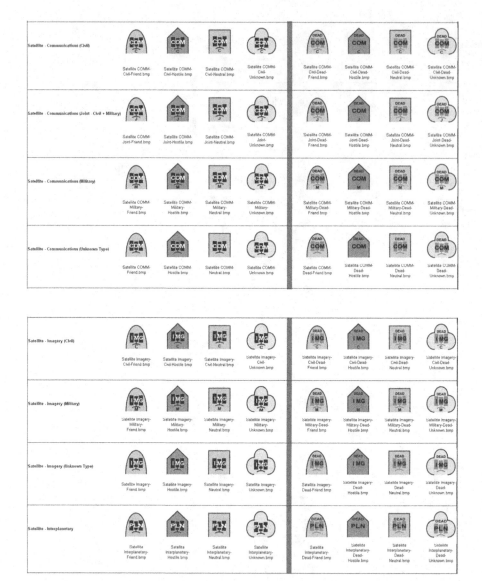

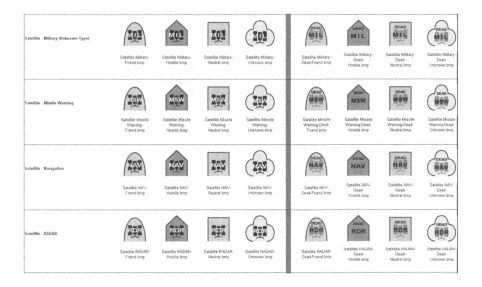

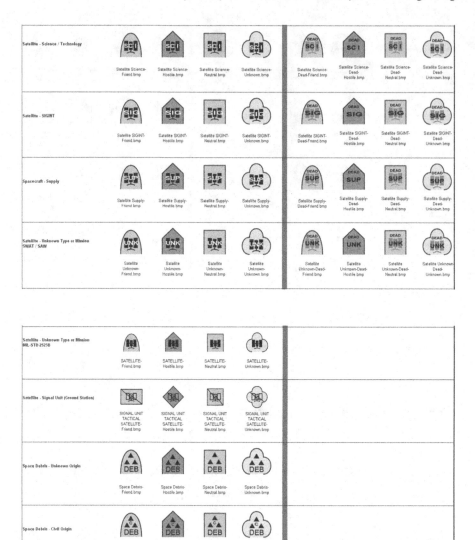

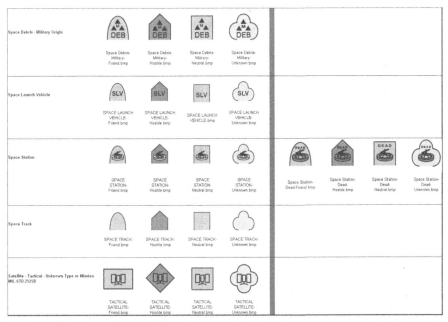

Last four sets of icons are from MIL-STD-2525B

REFERENCES

1. AFRL/RDTE, Satellite Assessment Center, SatAC Overview, "Satellite Assessments," Bill Mitchell, 23 July 2007.
2. Ibid.
3. Ibid.
4. Mr. Szymanski Corporation, "SHIVA - Space Highest Information Value Assessment Introduction," Paul Szymanski, 15 October 2008.
5. Department of the Air Force, "Counterspace Operations," Major Daniel D. Wright III, USAF, HQ AFDC/DR; Air Force Doctrine Document 2-2.1, 2 August 2004.
6. Department of Defense, "Dictionary of Military and Associated Terms," S. A. FRY, Vice Admiral, U.S. Navy, Director, Joint Staff; Joint Publication 1-02, 17 March 2009.
7. Belgian Astronomical Association, "Mike McCants' BWGS PPAS (Database of Photometric Periods of Artificial Satellites) Page," web site: http://www.io.com/~mmccants/bwgs/index.html, Mike McCants, 24 June 2009.
8. Mr. Szymanski Corporation, "SWAT Optical Data Analyses," Paul Szymanski, 17 June 2009.
9. Mr. Szymanski Corporation, "Choke Points Displays," Paul Szymanski, 12 September 2008.
10. Space.com, "U.S. Satellite Destroyed in Space Collision," Becky Iannotta and Tariq Malik, web site: http://www.space.com/news/090211-satellite-collision.html, 11 February 2009.
11. Department of Defense, "DOD Interface Standard - Common Warfighting Symbology," MIL-STD-2525B, 7 March 2007.
12. US Space Command, "JWID 00 Scenario," Maj. Dave Ehrhard, USSPACECOM JWID Team/J33T; 16 November 1999.

13. Space.com, "U.S. Satellite Destroyed in Space Collision," Becky Iannotta and Tariq Malik, web site: http://www.space.com/news/090211-satellite-collision.html (11 February 2009).
14. US Army Space and Missile Defense Center Public Affairs Fact Sheet: https://www.smdc.army.mil/Portals/38/Documents/Publications/Fact_Sheets/SWAT.pdf
15. The Design Knowledge Company, "Cognitive Task Analysis for the Joint Space Operations Center (JSpOC)," AFRL-RH-WP-TR-2009-00; March 2009.
16. Department of Defense, "DOD Interface Standard - Common Warfighting Symbology," MIL-STD-2525B, 7 March 2007.

10 Space and Information Analysis Model (SIAM)

Paul Szymanski

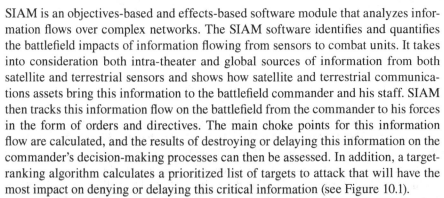

PART 1: SOFTWARE OVERALL DESCRIPTION

10.1 SIAM INTRODUCTION

SIAM is an objectives-based and effects-based software module that analyzes information flows over complex networks. The SIAM software identifies and quantifies the battlefield impacts of information flowing from sensors to combat units. It takes into consideration both intra-theater and global sources of information from both satellite and terrestrial sensors and shows how satellite and terrestrial communications assets bring this information to the battlefield commander and his staff. SIAM then tracks this information flow on the battlefield from the commander to his forces in the form of orders and directives. The main choke points for this information flow are calculated, and the results of destroying or delaying this information on the commander's decision-making processes can then be assessed. In addition, a target-ranking algorithm calculates a prioritized list of targets to attack that will have the most impact on denying or delaying this critical information (see Figure 10.1).

SIAM allows detailed analysis of information flows from information generation to intelligence processing, to command decisions, and finally to its effect on the battlefield. Most of the world's satellite sensors and networks have already been data-based within SIAM. Every step of the information flow can be analyzed for quality of information, survivability of processing elements, and timeliness. In addition, redundant paths can be assessed for their overall contribution to the information flow and survivability/resilience. The resulting analysis assists in identifying choke points, prioritizing targets, assessing weapons planning, illustrating the effects of including new sensor and C4 systems, and identifying intelligence collection shortfalls.

The SIAM tool runs under Microsoft Access. The original SIAM concepts were developed under the Air Force Research Lab sponsorship to show the value of advanced systems concepts to the warfighter. In addition to being fully tested in these exercises, it is supported with a complete user's manual. In addition, the SIAM software has been validated by the RAND Corporation in extensive testing.

10.1.1 DUAL USE

The SIAM software has a dual purpose: operational, and modeling/simulation. First, SIAM supports operational target prioritization of Red information flows. In this mode, a Candidate Target List (CTL) is developed as an input to the Aerospace Tasking Order (ATO) process. When applied to Blue information flows, vulnerabilities can be identified and the operational impact of outages can be assessed.

DOI: 10.1201/9781003321811-13

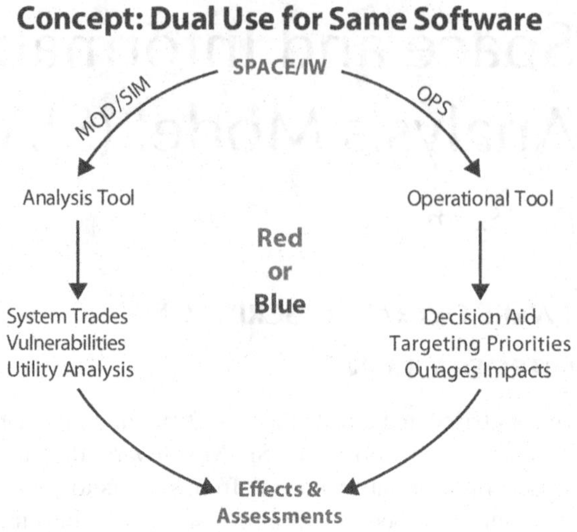

Concept: Dual Use for Same Software

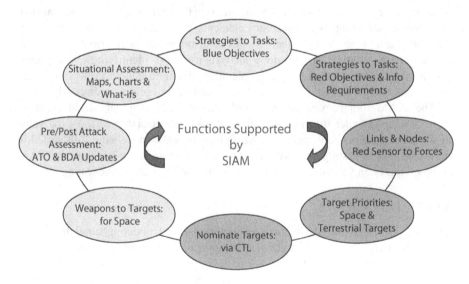

FIGURE 10.1   SIAM dual use.

Second, the modeling/simulation capability can show the utility of Blue weapons applied against Red information flows, or the utility of having Blue space systems in the context of a specific scenario. When applied to Blue C4I systems, this modeling and simulations capability can assess different system architectures.

SIAM's dual purpose is illustrated in Figure 10.1. The following descriptions focuse heavily on the operational target prioritization functionality of SIAM.

SIAM identifies and quantifies the battlefield impacts of information flowing from sensors to combat forces (see Figure 10.2). It takes into account both

FIGURE 10.2   SIAM process flow.

intra-theater and global sources of information. Both space-based and terrestrial sensors and communications are considered. The timeliness of information is the most critical criteria for assessing the impact of attacking selected nodes and paths.

10.1.2 ANALYSIS PROCESS

The SIAM analysis process significantly supports the assessment and planning functions of the Monitor-Assess-Plan-Execute (MAPE) targeting process (see Figure 10.3). In the ATO process, the cycle is repeated on a daily basis. SIAM outputs to the ATO planners include a prioritized CTL recommendation and supporting rationale in briefing format. Detailed screening of space-based satellites takes into account their orbital characteristics, sensor and site viewing windows, sunlight/night, and multiple other considerations. Targets can be assessed based on planned attacks and from Battle Damage Assessment (BDA) reports after they are attacked (see Figure 10.3).

10.1.3 POSSIBLE USES FOR SIAM

SIAM is an analytic tool that assesses information on the battlefield. It begins with military objectives and then tracks the flow of information on the battlefield from there. It can be used to

- Determine how information can affect battlefield decision-makers.
- Determine information choke points.
- Show results of denying or delaying this information.

SIAM analyzes a larger perspective on the battlefield.

- Considers both space and terrestrial resources
- Can analyze both red and blue perspectives

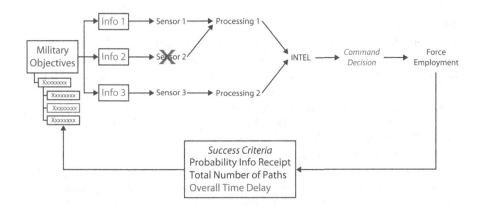

FIGURE 10.3   SIAM logic flow.

10.1.4 TECHNOLOGY HISTORY

10.1.4.1 Background

SIAM is a links and nodes analysis model that has been in development for over five years. The prototype tool is written with Microsoft Access databases. The original SIAM concepts were developed under the Air Force Research Lab sponsorship to show the value of advanced systems concepts to the warfighter. The Space Warfare Center saw the potential of SIAM to perform future planning analyses and sponsored the tool to play in future war games such as Blue-Flag and Joint Expeditionary Force Exercise (JEFX). In addition to being fully tested in these exercises, it is supported with a complete user's manual. The favorable warfighter's assessment of SIAM led to an HQ AC2ISRC/CC recommendation to CSAF that SIAM receive additional funding for full system development and subsequent fielding. This funding has led to the initiation of a complete rewrite of the tool in JAVA code using SQL server databases, using object-oriented software engineering and a spiral development process. It also led to a renaming of the tool to the Target Prioritization Tool for Links and Nodes (TPT-LN). SIAM v4.0 was the then-current TPT software plus additional export capability to support follow-up analysis.

10.1.4.2 Description

One of the original purposes of SIAM was to show the value of space to the warfighter. The value of space is sometimes hard to quantify since space assets provide mostly information in the form of imagery, missile warning, navigational signals, communications, and weather. SIAM assesses the value of this information by breaking down command decisions into requirements and then tracking how this required information is detected by sensors and how it flows through the battlefield from sensor to shooter (user). Specific information such as link reliability, link or node importance, timeliness, data rates, probability of receipt, etc., can be applied to each link and node throughout the system. SIAM can then assess the probability of information receipt, overall timeliness, and the quality of information received. Parametric analyses or "what if?" exercises can then be easily accomplished or new concepts can be added into the mix for a quick comparison analysis.

10.1.4.3 Uses

SIAM has been used in Red scenarios—what nodes (targets) or links (jamming, spoofing, etc.) can we eliminate to prevent the timely distribution of information to the Red commander? In Blue scenarios—what can we do to assure the Blue commander receives the correct information (vulnerability assessments) in time to support decision-making and the direction of forces? Or in strictly informational scenarios—calculate message delays through a system, or calculate overall system availability. In the past, SMC/Det 11 was employing SIAM to examine the space weather sites to input reliability and timing information to help determine where to place critical dollars for spares and to examine the effects of potentially eliminating sites to save resources. Another use could be in a test-planning scenario—to examine complicated systems to determine where to concentrate limited operational testing dollars and/or resources.

10.2   IDENTIFICATION AND SIGNIFICANCE OF THE PROBLEM

SIAM can measure the impact of key trends on important warfighting parameters. SIAM has an extensive database of all of the world's satellites and ground systems that are past, present, and planned for the future (see Table 10.1). The SIAM analyst can download differing scenarios for varying future time periods and country alignments. He can then give a detailed analysis of the impact on the battlefield of these space systems for each time period in question. This overall trending analysis shows the impact of new space systems of the future being employed by near-peer or third world countries that may be potential adversaries to the U.S. or may support those potential adversaries. At the same time, SIAM also includes terrestrial assets in its analyses, such as RPVs/UAVs, aircraft, ground radar, HUMINT, military patrols, forces, radio, microwave, fiber optic communications, intel centers, and command centers. This provides a comparison of space and ground methods of obtaining and transmitting information critical to battlefield decision makers (see Figure 10.3).

The growing availability of commercial imagery to support ISR functions is an increasing threat. SIAM assesses the deployment of new imagery satellites for the future. More importantly, it can track the flow of information along the connecting nodes that will handle the information from those commercial imagery satellites.

The increasing availability of commercial bandwidth-on-demand will support C2 and situational awareness by increasing the amount of information, sources of information, and timeliness of information. This increase in information flow will improve the C2 and situational awareness of the commander. The quality and timing of the information is important. The right information is needed in time to make a decision and act. SIAM can track the time and direction of information flow along the links and nodes of the information pathways.

Satellite navigation information can be tracked by SIAM, and the impact of this information on force decision-making can be assessed.

In each of these noted key trends, SIAM can identify the information pathways so that the command decisions and other required information can be targeted and denied or delayed. The important question is, did the commander receive the correct information in time to support his decision-making and direct his forces? If this can be prevented, then commercial imagery, commercial bandwidth-on-demand and the satellite navigation information, and the decisions based on this information can be affected.

TABLE 10.1
Nodes/Links in SIAM Database

Node Type	Total Number in Database
Satellites	7,630
Space-related Ground Sites	1,288
Other Ground Sites	3,098
Sensors	3,222
Node Totals	12,016
Link Totals	87,935

SIAM can quantify the changes to foreign and U.S. systems that rely on the increased use of space. Future satellites and space systems that can be identified and predicted can be loaded into the SIAM database. Any future date can then be selected, and then the space assets, communication links and nodes, and the information pathways available can be viewed. "What-ifs" can be constructed by turning selected space assets on or off, then viewing the resultant pathways, comparing them to terrestrial sources of information and communications paths. Non-space ground sensors and nodes (fiber optic cables, UAVs, etc.) can also be included in the future analyses. Scenarios showing the different levels and speeds of space buildup by foreign countries can be constructed, and the critical pathways can be identified and analyzed, along with providing technical threat assessments.

SIAM allows detailed analysis of information flows from information generation, to intelligence processing, to command decisions, and finally to its effect on the battlefield. Most of the world's satellite sensors and networks have already been databased within SIAM (see Table 10.1). Every step of the information flow can be analyzed for quality of information, survivability of processing elements, and timeliness. In addition, redundant paths can be assessed for their overall contribution to the information flow and survivability. The resulting analysis assists in identifying choke points, prioritizing targets, assessing weapons planning, illustrating the effects of including new sensor and C^4 systems, and identifying intelligence collection shortfalls.

10.3   INTERFACES TO WARGAMING MODELS

SIAM can provide the following outputs to wargaming models:

Total C4I degradation (total # of paths)
Total data bandwidth available
Probability of message receipt
Probability of alternative command decisions

SIAM would use the following inputs from wargaming models:

Forces probability of success
Nodes probability of survival

10.4   POSSIBLE SIAM ANALYSES

SIAM can conduct the following kinds of analyses:

Value of space
Value of C^4I systems
Info war
Space war
Weapon mixture comparisons
Target ranking
Future studies
Conflict level comparisons

C⁴I timeliness
Optimized attack type (deny/disrupt vs. degrade/destroy)
Threat periods
Intelligence collection prioritization
Space training

Benefits

Here are some of the benefits to using SIAM in the operational testing arena:

1. SIAM is a fully tested shrink-wrap product with a user's and training manual.
2. SIAM algorithms were independently verified and validated by the RAND Corporation in 1999 for SMC and was found to be a "good first step in developing a general IW attack and assessment capability."
3. SIAM has been used in five major war games at varying classification levels and received very high value ratings from the users.
4. SIAM can be interfaced to most external databases, and many databases are already built and available.
5. The SIAM software can automatically generate Microsoft PowerPoint charts to make briefings of the analysis performed.
6. The TPT software was independently verified, validated, and certified by the MITRE Corporation (acting as the government's representative).

PART 2: TYPES OF ANALYSES

10.5   SIAM DETAILED DESCRIPTIONS

SIAM is an objectives-based and effects-based software module that analyzes information flows over complex networks. The SIAM software identifies and quantifies the battlefield impacts of information flowing from sensors to combat units. It takes into consideration both intra-theater and global sources of information from both satellite and terrestrial sensors and shows how satellite and terrestrial communications assets bring this information to the battlefield commander and his staff. SIAM then tracks this information flow on the battlefield from the commander to his forces in the form of orders and directives. The main choke points for this information flow are calculated, and the results of destroying or delaying this information on the commander's decision-making processes can then be assessed. In addition, a target-ranking algorithm calculates a prioritized list of targets to attack that will have the most impact on denying or delaying this critical information (see Figure 10.1).

A description of the SIAM main software screens follows.

Figure 10.4 shows the main SIAM screen when the software is first started. We will now discuss individual functions when clicking on the specific buttons.

10.5.1   Scenario Generation (Country Alignments)

This screen (Figure 10.5) allows the analyst to list which countries are on the Red, Blue, and Gray (contributes to both sides) sides of the scenario being analyzed. Choose from the drop-down menu, or enter a new country at the bottom of the list.

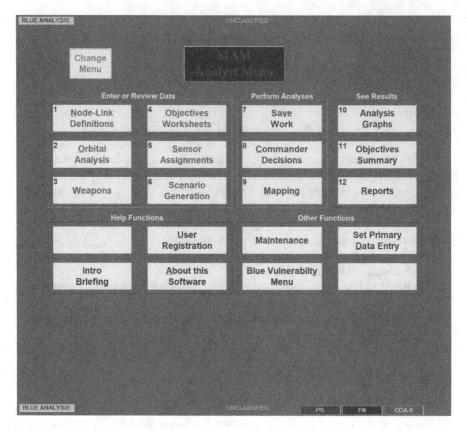

FIGURE 10.4 SIAM main screen.

Delete the countries on each list (side) that are not to be included in the scenario. Do this by putting the cursor in the field of the country that you want to delete and then clicking the delete button on your keyboard. The Blue and Red sides both must have at least one entry in them.

Some countries (France, United Kingdom, United States) have more than one entry because of alliances and treaties. The Gray side contains several satellite consortiums that are made up of two or more countries linked for business purposes. This consortium may act differently and independently from the countries that belong to it.

10.5.2 COMMANDER DECISIONS

Figure 10.6 shows the commander's decisions supporting the military objectives in the top portion, and the corresponding information requirements for each command decision in the bottom portion (from the information requirements worksheet screen). It also allows weapon attacks against this information.

The information requirements are broken down into four categories: Opposing Force Status, Own Force Status, Terrain Status, and Weather Status. Each category

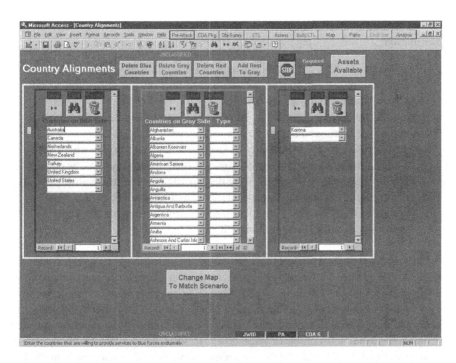

FIGURE 10.5   Country alignments user screen.

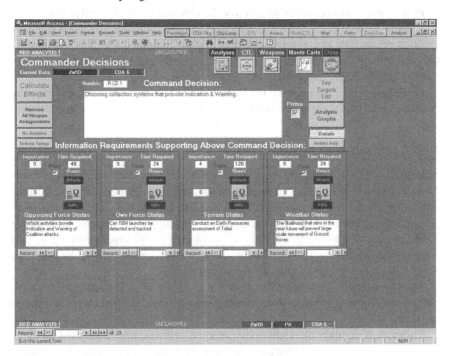

FIGURE 10.6   Commander's required decisions.

lists an information requirement and shows the following characteristics of that information:

- Importance—Priority of the information-collection task to be completed as compared among the four information requirements categories (on a scale of 1 to 10, with 10 the highest).
- Time Required (in Hours)—Lead time needed to have the information to support the command decision prior to the event.
- Minimal Acceptable Probability of Completion—Determines which paths will be considered in the analysis (0 means that all pathways will be considered for the analysis; 90 means only the top 10 percent of the pathways will be considered). 0 is the default value; 1–99 can be entered.
- Attack Info—Takes you to the Paths screen where weapons can be applied against nodes in the information pathway.

Each record will have one combination of the five items made up by the command decision (one item) and information requirements (four items). Only one of these combinations (records) can be analyzed at a time. You can also click on the drop-down arrow within each information requirement window to see other information requirements associated with that command decision. Scroll through the records using the arrow in the lower left corner, or page down, to view the possible combinations of command decision and information requirements (if more than one combination has been constructed).

Prime checkbox is checked to identify the commander decision that is currently being used for analyses when there is more than one commander decision (record).

The Commander Decisions screen displays the command decision and information requirements that are input on the Information Requirements Worksheet screen (Objectives Worksheets button) at the Analyst Menu.

A pathway may take up more than one record. The blue "Begin Path" identifies the node at the beginning of the path, and the red "End Path" identifies the last node.

10.5.3 ATTACK INFO (…PATHS)

This button on the Command Decisions screen leads to the Paths screen (Figure 10.7) for each of the four information requirement categories. This displays the information pathways for the scenario. The path number is displayed in the center of the screen, and the total number of paths is shown in the upper right corner of the screen. The sequence of flow is shown from the sensor, to the receiver terminal, through communications, to the INTEL Center, through communications, to the command center, through communications, to the military forces.

The Opposing Force Info screen is shown in Figure 10.7. The screens for each of the four categories (Opposing Force, Own Force, Terrain, and Weather) are designed the same.

This screen displays the eight nodes on each pathway, and it is sorted automatically to show the most probable path first (but the number of this path will probably not be 1). If there are more than eight nodes in the pathway it will be continued on

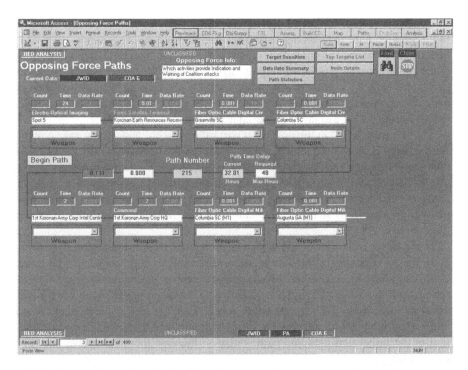

FIGURE 10.7   Information paths.

the next record. The Begin Path box denotes the start of the path, and the End Path box identifies the end of the path.

There are two basic ways to disrupt the information flow. The user could attack the most probable path first, by looking for the biggest choke point on that path. The number of pathways in which the node appears can be seen above the node name in the "Count" field. A larger number indicates that node is a bigger choke point and, therefore, eliminates the most paths if it is taken out of the network. Alternatively, the user can attack nodes based on the prioritized target list shown on the Top Target screen and the Prioritize CTL screen.

The list of paths can also be sorted by time, and the user can then attack the most timely paths. To sort, place the cursor in the Path Time Delay Current Hours field and click on the sort A→Z button on the access toolbar.

The selection of attack type is based on the strategy or tactics the user may want to employ, derived from the CINC's intent. Select a weapon from the drop-down menu under the node that you want to target. For strategic targeting, it is best to select just one node to target in one pathway of the information requirement (selected from one of the four categories, for one command decision). Selecting multiple weapons laydowns for one path might be wasteful, as destroying one node on the path may be the only attack necessary to eliminate that path, and possibly more paths. One node will probably show up in several pathways, across many information requirements (collateral damage). To view a new record, use the arrows in the lower left corner, or page down.

The Paths screens will display different pathways for each of the four categories of information requirements discussed, because they are probably supported by different sensors (as well as other nodes). However, it is common for different information requirements to use the same pathways, or to use the same "communications backbone" (where the main or common part of a pathway is the same with different nodes at one or both ends). The opposing force category was selected as an example, but all categories are displayed in the same format.

The "Current Data" in the upper left part of the screen shows the scenario name (in the black box) and the analysis short name (in the brown box) for the analysis that is displayed on the Paths screen. The paths are a cumulative summary of the effects that attacking certain targets has on the pathways, so as soon as another analysis has been run, this screen will change. You cannot go back and view the paths from a previous analysis. You can only view the paths from the last analysis that was run.

In the lower right part of the screen is shown the scenario name (in the purple box), the ATO phase (in the blue box) and the analysis short name (in the green box). The analysis named here is the one selected on the Set or Delete Analysis screen, and the Target Densities and Path Statistics button (dialogue boxes) will display those results. When you name a new analysis, that analysis short name will appear here before you conduct the calculation, and there will be no information in the target densities and path statistics (because the results have not been calculated).

The Target Densities button brings up the Target Densities dialog box. This graphically shows the number of targets for each of the eight types of nodes.

The Data Rate Summary button brings up the Data Rates dialog box. This graphically shows the data rates for each of the eight types of nodes.

The Sensor-Rcvr Info Paths button goes to the Sensor-Receiver Info Paths screen. It shows the different sensor to receiver parts of all of the paths. This can be useful in analyzing the similarities and differences of the pathways.

The Analysis Graphs button goes to the Analysis Graphs screen.

The Details button is usually on, which allows you to view all of buttons normally on the Command Decisions screen. When this button is clicked off, only the CTL and Close/Stop buttons are viewable and available for use.

The Delete Info button deletes the tables that contain the information to display the paths associated with the four information types. It may be useful to delete these tables when you want a minimum file size when shipping a copy of SIAM. It may also be useful when a large analysis run is about to be conducted that might cause the size of the database to go over the 2 gigabyte file size limit (after clicking the button also compact the database). After this button is clicked you will have to recalculate the results again before you can see the paths displayed.

The Attack Info button goes to the Paths screens where you can view the information pathways for that information category.

The Prime checkbox in the information requirements section is checked to identify the information requirement that is currently being used for analyses when there is more than one information requirement in one of the categories (Opposing Force, Own Force, Terrain, Weather).

10.5.4 Top Targets List (Top Targets)

Figure 10.8 shows the target ranking in a tabular display, as a result of the last analysis conducted, according to their relative score.

The Probability of Success column on the right of the screen comes from that same field on the Node Date Entry screen.

The Top Targets Graph button goes to the Top Targets – Current Analysis screen (Figures 10.9 and 10.10). It shows a graph of the top targets listed from left to right in order of their relative score.

The Sort buttons allow you to sort the nodes by score (high to low); count (the total number of choke points for that node) (high to low); probability of success (high to low); time delay (low to high); and data rate (high to low). Click on the sort button over the Score column to return to the original order.

10.5.5 Strategies (Attack Strategy)

Intent goes to the CINC Direction and Intent screen (Figure 10.11). This lists the CINC intent to be used in the scenario along with a description for each. The scenario name and short name are listed at the top.

Figure 10.12 lists the strategies to be used in the scenario along with a description for each. The scenario name and short name are listed at the top.

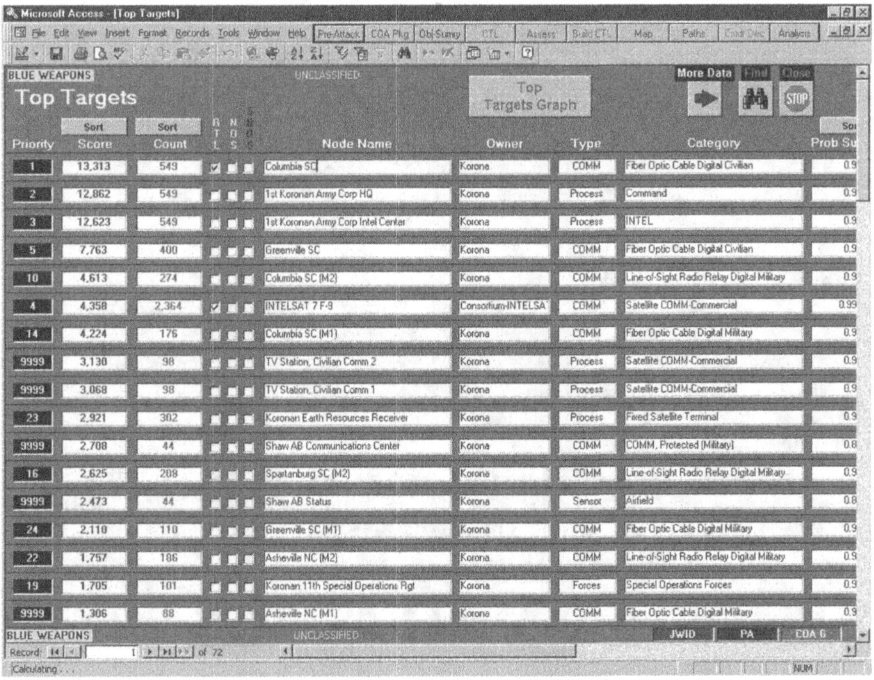

FIGURE 10.8   Top battlefield targets (both space and terrestrial).

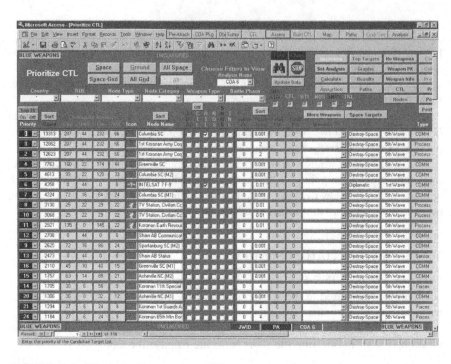

FIGURE 10.9 Weapon assignments attacking network nodes.

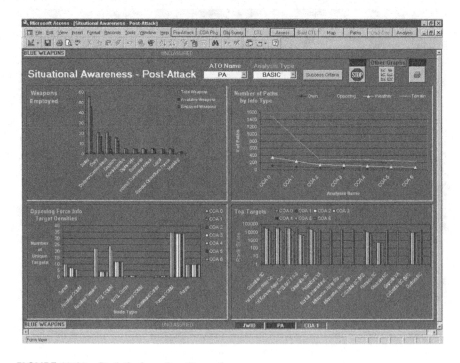

FIGURE 10.10 Statistical results of targeting.

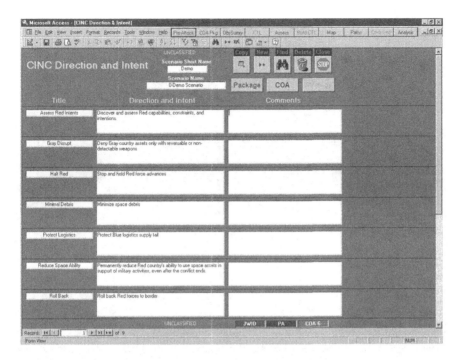

FIGURE 10.11 Senior leadership directions and intents.

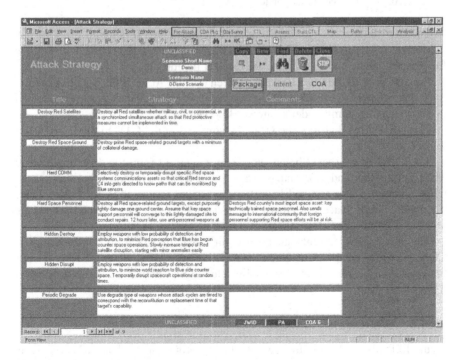

FIGURE 10.12 Definitions of overall attack strategies.

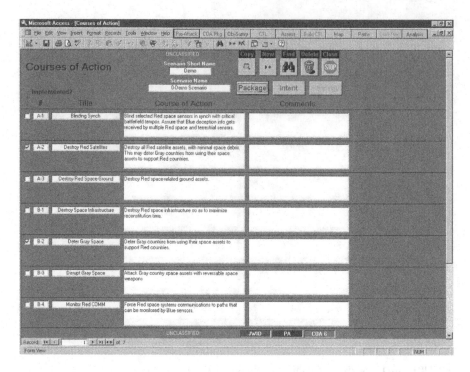

FIGURE 10.13   Selected courses of action.

The COA button goes to the Courses of Action screen (Figure 10.13). This lists the courses of action to be used in the scenario along with a description for each. The full scenario name and short name are listed at the top.

Package goes to the COA Packages screen (Figure 10.14). This is a bookkeeping form that is used to keep track of the course of action packages that are assembled to carry out the commander's intent. However, it also shows the strategies-to–tasks relationships that link the targets nominated to the CTL to the CINC intent (objectives).

The CINC intent that is being supported is selected from the drop-down menu. The COA number and COA name, strategy and weapons laydown that support that CINC intent are also selected from the drop-down menus at each of those fields. Then the CTL, or Space STL, that was constructed to implement that CINC intent is selected from the drop-down menu.

The same CINC intent can be accomplished by different combinations of COA, strategy, weapons laydown, and CTL/Space CTL. These different combinations can be ranked in priority by selecting a number from the drop-down menu.

10.5.6   Weapon Definition

Figure 10.15 allows the user to view/enter the types of weapons that can be used in the scenario. Drop-down menus are provided for the following characteristics: Weapon

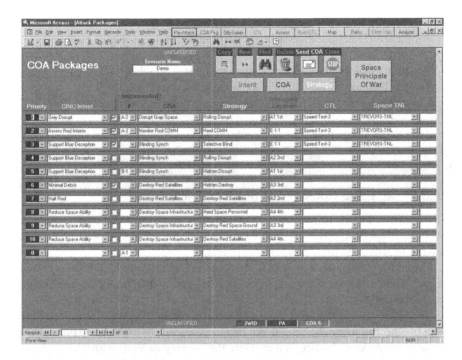

FIGURE 10.14   Overall courses of action packages.

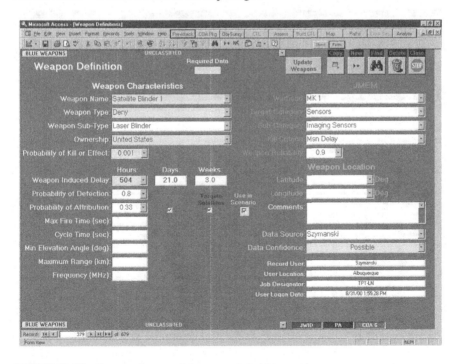

FIGURE 10.15   Space and terrestrial weapon definitions screen.

Name, Weapon Type, Ownership, Probability of Kill Effect, Weapon Induced Delay, Probability of Detection, Probability of Attribution, and Comments.

Weapon Available box: Check if the weapon is available for use.
Targets Satellites box: Check if the weapon can target satellites.
Use In Scenario box: Check if this weapon is to be used in this scenario.

After a new weapon has been entered, or changes have been made to this screen, click on the Update Weapons button to update the database.

10.5.7 ROE AND COUNTRY LIST (RULES OF ENGAGEMENT)

The screen in Figure 10.16 (column 8) allows the user to pair the country involved with the Highest Allowed Weapon category (ROE) that can be used against that country by using the drop-down menus provided.

The country codes used by NASA, the SATCAT, NAIC, and MIDB are also listed because they differ for some countries. This is useful when importing information from different databases.

The country or ROE can be sorted by selecting one of the fields at the top of the page in blue and using the drop-down menus provided for each one. After entering a selection, click on the See button. Only the country or ROE with that entry will

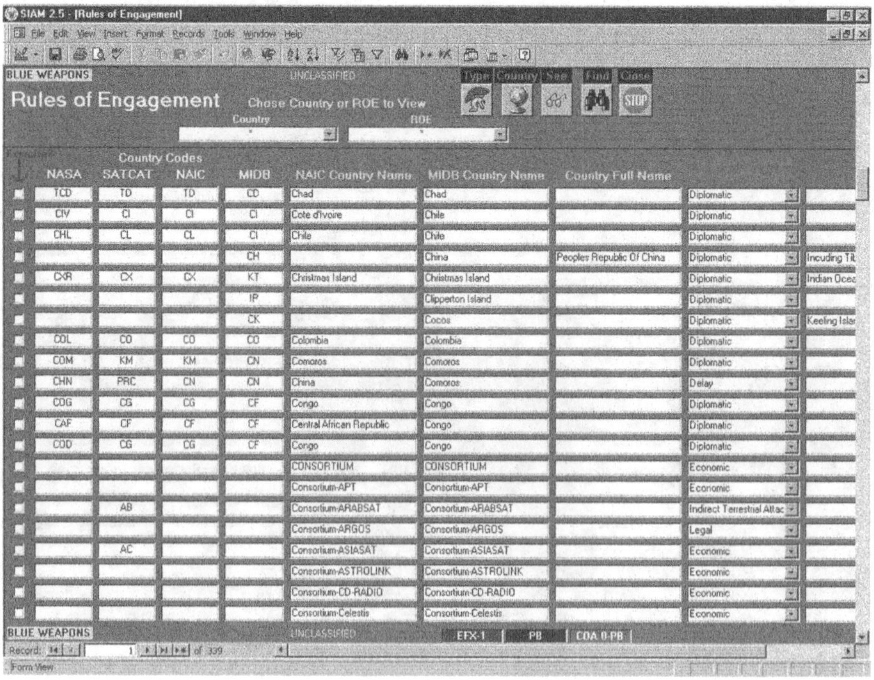

FIGURE 10.16 Rules of engagement.

be shown. To restore the screen to the original condition showing all of the records, place a " * " back in the blue sort fields and click on the See button.

10.5.8 OBJECTIVES WORKSHEETS (MILITARY OBJECTIVES)

Figure 10.17 shows a menu screen illustrating the military objectives for the Blue and Red sides. It shows who (Blue or Red) is considered "Own Side" and who (Blue or Red) is considered "Opposing Side." This is based on whether you are doing a Blue or Red analysis (selected on the Assets Available screen). Click on the buttons to go to the respective screens. Explanations of these screens follow.

At the box in the upper left corner of the screen you can select the priority of the objectives that you want to view. The drop-down lets you choose a number that corresponds to the lowest priority of objectives that will appear on the Objectives Summary screen.

This screen shows the objectives from the national goals down to the command decision for both Blue and Red. The category, type and number of each objective is shown. The number of the record showing is at the top center of the screen just under the Objectives Flow Name field. Directly under that box is the Link Objective column. This is used to record the level at which the Blue and Red objectives intersect, or directly affect, each other. Place a check in the box at the corresponding level and a black line will link the Blue and Red objectives at that level.

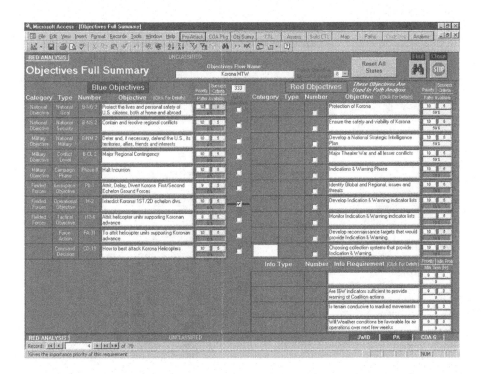

FIGURE 10.17 Military objectives.

To view different records on the Objectives Full Summary screen:

Blue Objectives: Use the forward and back buttons associated with the Access
   Record counter in the lower left corner of the screen. It will show you the
   current record that is displayed. This record number is also shown at the top
   center of the screen.
Red Objectives: Use the Page Up/Page Down buttons on the keyboard. First you
   must click in any box/field on the Red side. This will activate to the Page Up/
   Page Down buttons to work with the Red records. (The Page Up/Page Down
   buttons will also work to view the Blue records if you have clicked on that side.
   However, this is the only method that allows viewing of the Red records.)

The information requirements corresponding to the command decision are shown
at the bottom of the screen. The inputs for the Priority, Min Prob, and Min Time
(Hr.) fields are made on the Information Requirements Worksheet screen.

10.5.9   VIEW OBJECTIVES VS. TARGETS (OPPOSING SIDE OBJECTIVES VS. TARGETS)

Figure 10.18 shows the connection between the current objectives tree and the targets
that have been attacked so far in the analyses. The objectives shown are at the same
level in the tree as the Select Objectives Level For Graph box (on the Objectives Full
Summary screen).

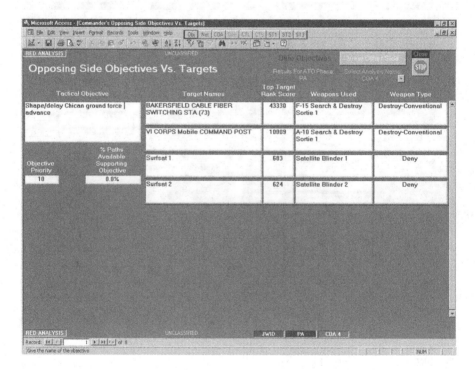

FIGURE 10.18   Red objectives vs. selected targets.

Each screen view shows one of the objectives at that level and all of the targets that have been attacked in support of that objective. It shows the Blue side objectives (if you are doing a Red side analysis). It also shows other information, such as the weapon and weapon type used for the attack.

Use the Page Up and Page Down keys to see the other objectives that are affected.

The View Other Side button takes you to the Commander's Own Side Objectives Vs Targets screen. It allows you to see how the Red side objectives are being affected by the targets being attacked. It shows the same information as the <u>Opposing Side Objectives Vs Targets</u> screen.

10.5.10 Graph Rank Order vs. ATO (Objectives Rank Order of Importance vs. ATO Phase)

Figure 10.19 is the actual graph of objective priority over time. The left side of the screen lists all of the objectives at the level being tracked. They are in the rank order, as entered, for the first ATO phase. Their rank order is also shown. The graph displays lines of different colors, one for each objective from the left side list.

10.5.11 Report Selected Objectives in Tree Format (Objectives Tree Report)

Figure 10.20 shows the print preview mode for the tree report. It shows all of the objectives in the selected tree, in an indented format to show which ones are grouped below each higher-level objective.

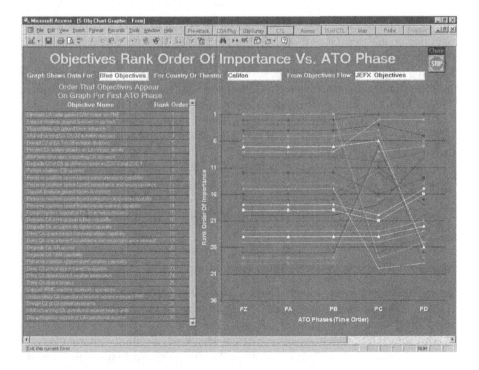

FIGURE 10.19   Military objectives rankings by battle phase.

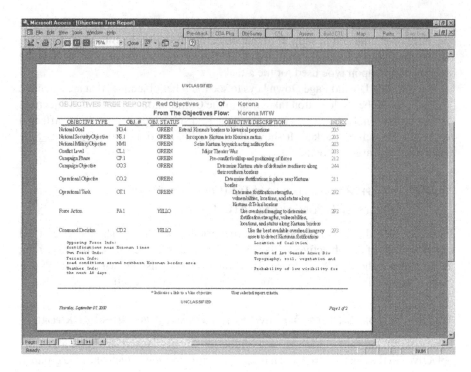

FIGURE 10.20　Military objectives report.

The combo boxes at the top of this screen allow you to select a particular objectives tree or sub-tree. Once those selections are made, you can then see all of those objectives in an outline or tree format on this screen, or you can print a copy.

If you tab through or change the selection in the Select One Type of Objective or Info Requirement field, you then get a Print Only Selected Type of Objectives button. This will print a list of all of the objectives or info requirements that you selected.

For example, if you have selected opposing force information, you will see all of those information requirements that are in the tree for the selected country regardless of where they fall in the tree.

10.5.12　SENSOR ASSIGNMENTS

Figure 10.21 is a menu screen for sensor assignments. A discussion of the buttons on this screen follows.

10.5.13　ASSIGN SENSORS TO TARGETS

Figure 10.21 gives a description of each sensor and its capabilities. The fields can be filled in using the drop-down menus. Fields in blue are required data and must be filled in.

This screen lists which sensors can detect what kind of military information.

The Nothing Assigned button goes to the Sensors With No Assignments screen.

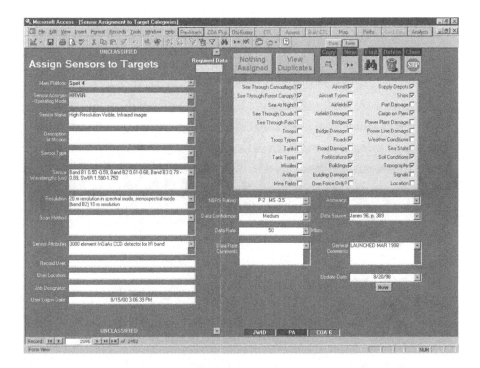

FIGURE 10.21   Assigning sensor to targets.

The View Duplicates button goes to the Assign Sensors to Targets-Duplicates screen. The Now button under the Update Date field will put in the current date.

10.5.14   ASSIGN OBJECTIVES TO TARGETS

Figure 10.22 lists the information requirements for the categories of Opposing Forces, Own Forces, Terrain, and Weather. It then gives the detection capabilities needed to determine those information requirements.

The numbers are an indication of the relative importance (on a scale of 0 to 10) of that particular capability that is needed to determine the information required. If it is not known what capability is required for a particular information requirement, you could put a 5, for example, in each capability field to equalize the relative importance. The box below the number must be checked if it is to be used in the analysis.

10.5.15   SENSOR ASSIGNMENT SUMMARY

Figure 10.23 gives a summary by category (Opposing Force, Own Force, Terrain, Weather) and information requirements of the sensor that can fulfill the requirement to detect the objective. It also gives the degree of requirement fulfillment (how well that sensor satisfies the information requirement, with 1 being 100 percent satisfied) and the data confidence.

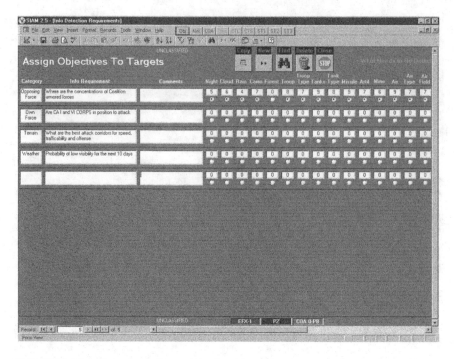

FIGURE 10.22   Assigning military objectives to targets.

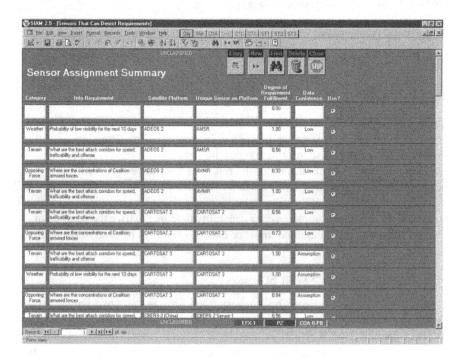

FIGURE 10.23   Sensor assessments summary.

10.5.16 Review Best Sensor Assignments

Figure 10.24 lists the sensor assignments by information type (Own Force, Opposing Force, Terrain, Weather), information requirement, what the sensor is flown on, and the relative value of the information.

This summary screen is a data sheet view of the sensor assignments automatically generated by SIAM. It lists the sensor assignments by information type (Own Force, Opposing Force, Terrain, Weather), information requirement, what sensor platform the sensor is on, and the relative INTEL value of the information. The INTEL value is expressed as a percent, and it shows the amount of the types of information listed on the Assign Objectives to Targets screen that are answered by a sensor.

10.5.17 See Analysis Graphs (Analysis Graphs)

This button leads to the Analysis Graphs screen. For a detailed explanation see the Analysis Graphs topic under the See Results section of this chapter (see "Analysis Graphs" button in Figure 10.4 that leads to Figure 10.10).

10.5.18 Details

This button (Figure 10.6) is normally in the clicked position. If you unclick this button you get a summary of the commander decisions screen. This is meant to simplify the screen to show only the key information to a less experienced user. It

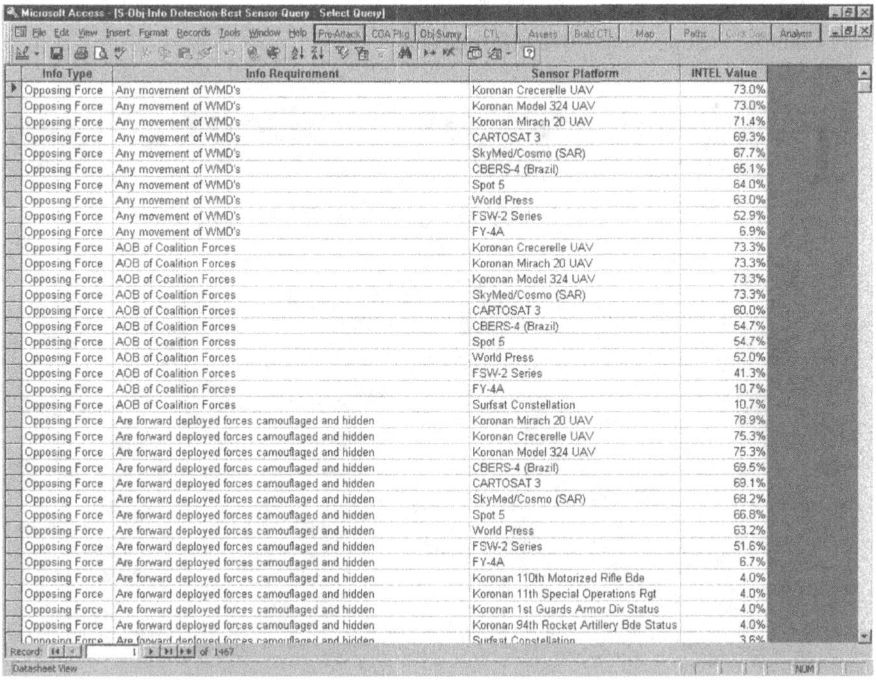
Microsoft Access - [S-Obj Info Detection Best Sensor Query : Select Query]

Info Type	Info Requirement	Sensor Platform	INTEL Value
Opposing Force	Any movement of WMD's	Koronan Crecerelle UAV	73.0%
Opposing Force	Any movement of WMD's	Koronan Model 324 UAV	73.0%
Opposing Force	Any movement of WMD's	Koronan Mirach 20 UAV	71.4%
Opposing Force	Any movement of WMD's	CARTOSAT 3	69.3%
Opposing Force	Any movement of WMD's	SkyMed/Cosmo (SAR)	67.7%
Opposing Force	Any movement of WMD's	CBERS-4 (Brazil)	65.1%
Opposing Force	Any movement of WMD's	Spot 5	64.0%
Opposing Force	Any movement of WMD's	World Press	63.0%
Opposing Force	Any movement of WMD's	FSW-2 Series	52.9%
Opposing Force	Any movement of WMD's	FY-4A	6.9%
Opposing Force	AOB of Coalition Forces	Koronan Crecerelle UAV	73.3%
Opposing Force	AOB of Coalition Forces	Koronan Mirach 20 UAV	73.3%
Opposing Force	AOB of Coalition Forces	Koronan Model 324 UAV	73.3%
Opposing Force	AOB of Coalition Forces	SkyMed/Cosmo (SAR)	73.3%
Opposing Force	AOB of Coalition Forces	CARTOSAT 3	60.0%
Opposing Force	AOB of Coalition Forces	CBERS-4 (Brazil)	54.7%
Opposing Force	AOB of Coalition Forces	Spot 5	54.7%
Opposing Force	AOB of Coalition Forces	World Press	52.0%
Opposing Force	AOB of Coalition Forces	FSW-2 Series	41.3%
Opposing Force	AOB of Coalition Forces	FY-4A	10.7%
Opposing Force	AOB of Coalition Forces	Surfsat Constellation	10.7%
Opposing Force	Are forward deployed forces camouflaged and hidden	Koronan Mirach 20 UAV	78.9%
Opposing Force	Are forward deployed forces camouflaged and hidden	Koronan Crecerelle UAV	75.3%
Opposing Force	Are forward deployed forces camouflaged and hidden	Koronan Model 324 UAV	75.3%
Opposing Force	Are forward deployed forces camouflaged and hidden	CBERS-4 (Brazil)	69.5%
Opposing Force	Are forward deployed forces camouflaged and hidden	CARTOSAT 3	69.1%
Opposing Force	Are forward deployed forces camouflaged and hidden	SkyMed/Cosmo (SAR)	68.2%
Opposing Force	Are forward deployed forces camouflaged and hidden	Spot 5	66.8%
Opposing Force	Are forward deployed forces camouflaged and hidden	World Press	63.2%
Opposing Force	Are forward deployed forces camouflaged and hidden	FSW-2 Series	51.6%
Opposing Force	Are forward deployed forces camouflaged and hidden	FY-4A	6.7%
Opposing Force	Are forward deployed forces camouflaged and hidden	Koronan 110th Motorized Rifle Bde	4.0%
Opposing Force	Are forward deployed forces camouflaged and hidden	Koronan 11th Special Operations Rgt	4.0%
Opposing Force	Are forward deployed forces camouflaged and hidden	Koronan 1st Guards Armor Div Status	4.0%
Opposing Force	Are forward deployed forces camouflaged and hidden	Koronan 94th Rocket Artillery Bde Status	4.0%
Opposing Force	Are forward deployed forces camouflaged and hidden	Surfsat Constellation	3.6%

Record: 1 of 1467

Datasheet View

FIGURE 10.24 Sensors INTEL value assessment.

shows the command decision and the four information requirements (including their importance, time required in hours, and minimum acceptable probability of completion). For a full explanation, see Decisions under Prioritize CTL (Candidate Target List Assessment, Figures 10.47 and 10.48) in the Analysis section of the Operations Menu Functions chapter.

10.5.19 TARGET DENSITIES

The Target Densities button brings up the Target Densities dialog box (Figure 10.25). This graphically shows the number of targets for each of the eight types of nodes.

Double-click "Name" to see path densities: Double-click on the type of target (across the bottom) and another dialog box will appear that shows the individual nodes of that type that are behind the numbers on the graph.

10.5.20 DATA RATE SUMMARY

The Data Rate Summary button brings up the Data Rates dialog box (Figure 10.26). This graphically shows the data rates for each of the eight types of nodes.

10.5.21 PATH STATISTICS

The **Path Statistics** button brings up the Path Statistics dialog box (Figure 10.27). This shows the time statistics totals for all of the different paths in the category, in this example the opposing force information.

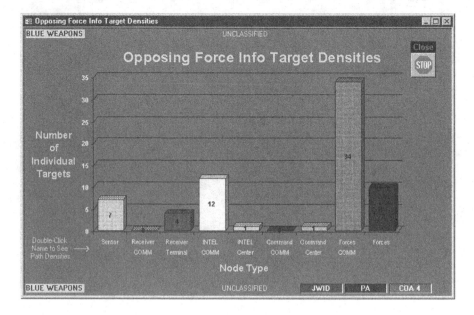

FIGURE 10.25   Target densities graph.

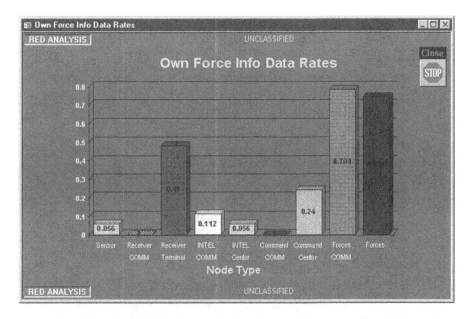

FIGURE 10.26 Data rate summary graph.

10.5.22 MAPPING

This leads to the map display in SIAM (Figures 10.28 and 10.29). The type of map that appears is the map that is set as the default map.

The gray area surrounding the map contains any nodes that are selected for viewing but do not appear on colored portions of the map. This will include any satellites in space, as well as nodes that may no longer appear on the map after you zoom in.

The Map Toolbar and the Map Functions Buttons toolbar are explained in the following topics.

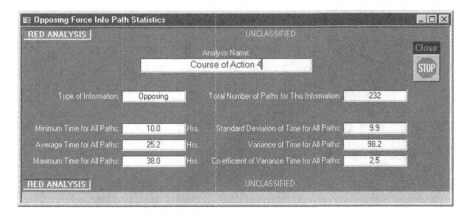

FIGURE 10.27 Path statistics graph.

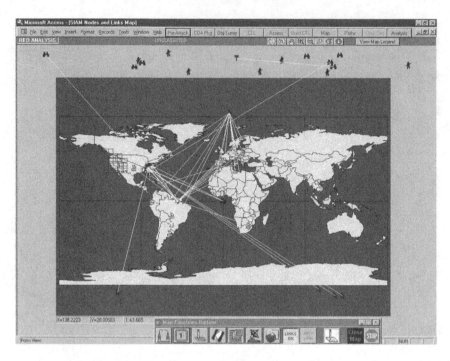

FIGURE 10.28   Information flow links map—overview.

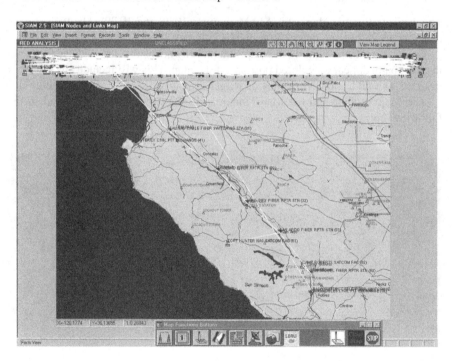

FIGURE 10.29   Information flow links map—detail.

Choke Point Graphs—This screen has the following four options:

Opposing Force Choke Points—Lists the chosen node types and the opposing force information choke points.

Own Force Choke Points—Lists the chosen node types and the own force information choke points.

Terrain Choke Points—Lists the chosen node types and the terrain info choke points.

Weather Choke Points—Lists the chosen node types and the weather information choke points.

The four Choke Point screens are displayed in the same format. An example of the format is shown in Figure 10.30.

Top Node Graph—Figure 10.31 gives the nodes of the chosen node type and the rank score by each analysis name.

10.5.23   ALL PATHS PROBABILITY GRAPH

Figure 10.32 shows the overall probability of the information getting through by any path, for each analysis conducted. It must be used in conjunction with the Monte Carlo Settings screen (Monte Carlo button on the Command Decisions screen).

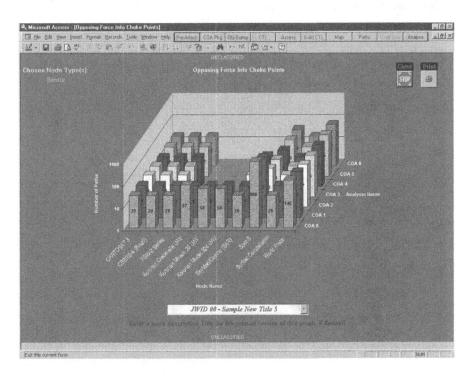

FIGURE 10.30   Top nodes as choke points graph.

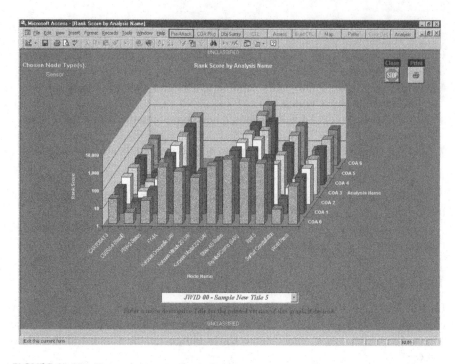

FIGURE 10.31   Top nodes to attack graph.

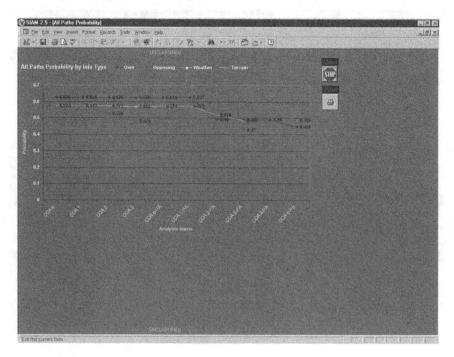

FIGURE 10.32   All paths probability graph.

Time Graphs screen (Figure 10.33) has the following seven options:

Maximum
Average
Minimum
Variance
Standard deviation
Coefficient of variance
Time distributions (time distribution graphs)

The analysis type to be displayed will be shown in the Choose Analysis Type section. A different analysis type can be selected from the drop-down menu. This analysis type is from the green box on the Set or Delete Analysis screen. Two analysis types can be viewed on a graph at one time by selecting a second type from the drop-down menu of the blank field below the first.

The seven time screens are displayed in the same format. An example of the format is shown in Figure 10.33.

Time distributions (time distribution graphs) of the time graphs in Figure 10.33 has the following four options:

Opposing force time distribution
Own force time distribution

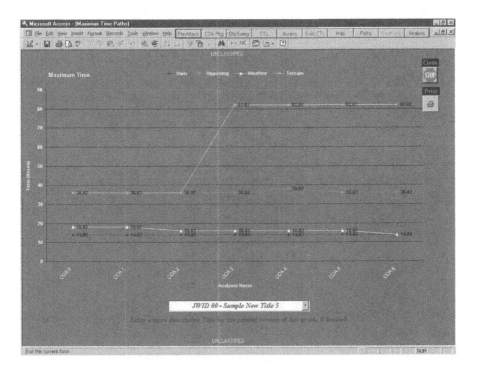

FIGURE 10.33   Information paths time delays graph.

Terrain information time distribution
Weather information time distribution

These graphs show the number of paths and total delay time for each of the Info Requirement categories. It allows you to make comparisons among the analysis runs.

Enter the "Selected Analysis Type" in the designated field, then click on the button desired.

The four time distribution screens are displayed in the same format. An example of the format is shown in Figure 10.34.

The Data Rate Graphs screen has the following nine options:

Sensors
Receiver-COMM
Receiver terminals
INTEL-COMM
INTEL Centers
Command centers COMM
Forces
Forces COMM
Command centers

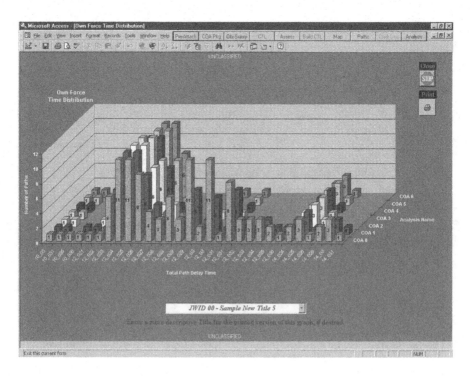

FIGURE 10.34   Information paths time delays graph.

The nine data rate graphs are displayed in the same format. An example of the format is shown in Figure 10.35.

The Weapon Graphs screen has the following 11 options:

Probability of detection graph
Probability of attribution graph
Weapon delay by weapon type
Weapon delay by analysis name graph
Weapon use by type graph
Weapon use by sub-type graph
Weapon use by owner graph
Weapon use by reliability graph
Weapon use by target category graph
Weapon use by target sub-category graph
Weapon use by target kill criteria graph

The analysis type to be displayed will be shown in the "Choose Analysis Type" section. A different analysis type can be selected from the drop-down menu. This analysis type is from the green box on the Set or Delete Analysis screen. Two analysis types can be viewed on a graph at one time by selecting a second type from the drop-down menu of the blank field below the first.

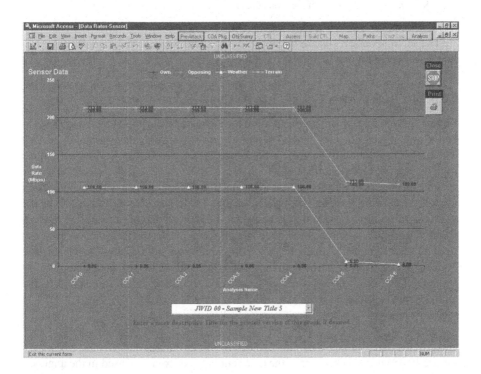

FIGURE 10.35   Information paths data rates graph.

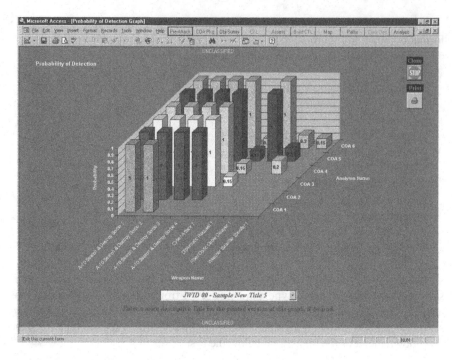

FIGURE 10.36   Probability of weapons use detection graph.

The 11 weapon screens are displayed in the same format. An example of the format is shown in Figure 10.36.

10.5.24   TOP TARGETS GRAPH (TOP TARGETS—CURRENT ANALYSIS)

Figure 10.37 is a graphical representation for the nodes determined by the analysis to be the top targets. They are ranked in order by a relative score determined by SIAM. The higher score directly relates to the value of the target. It can be used to identify high value nodes for targeting.

The targets reorder themselves after each analysis (after a weapon has been applied) according to their new relative score. The relative score is a result of the top node ranking algorithm that is based on the importance of information flowing through the node, how much the node is a choke point for the information flow, the total number of nodes of that category, how capable the node is, how survivable the node is, the probability of the path the node is on, and the overall attack strategy.

Note: The Top Targets screen will show the top targets that correspond to the analysis that is selected on the Set or Delete Analysis screen.

10.5.25   WEAPON TYPES USED

Figure 10.38 shows a pie chart of the different types of weapons used in the analyses by percent.

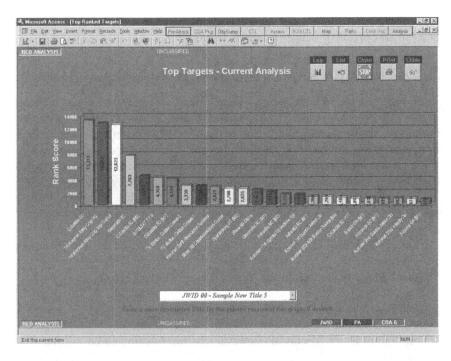

FIGURE 10.37   Recommended top network nodes to attack.

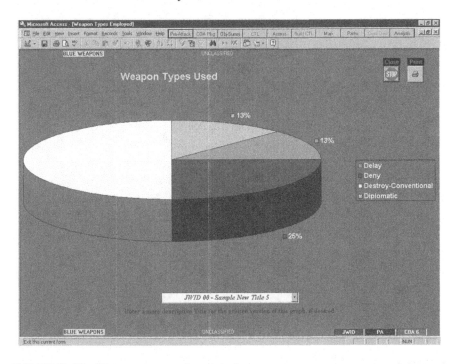

FIGURE 10.38   Weapon types employed graph.

See the Weapon Types topic under the Weapons section of Enter or Review Data for a full description of weapon types.

10.5.26 Pre-Attack Situation (Situational Awareness—Pre-Attack)

Figure 10.39 lets you view the current situation as represented by the state of the analyses that have been conducted. This is intended to show the situation after only a baseline analysis has been conducted, as part of the new SIAM-loaded database. However, it will also update and show the situation after each additional analysis has been run.

Different screens can be viewed, sorted by ATO name and analysis type. Click on the drop-down arrow in these fields to select another name or type to view.

The screen is divided into four sections: Weapons Available, Number of Paths by Info Type, Opposing Force Info Target Densities, and Top Targets. Click on one of the sections to enlarge that screen and see more details.

Each of these four screens can also be sorted and viewed by ATO name and analysis type.

Other buttons on the Situational Awareness—Pre-Attack screen:

Close/Stop button returns to the Analysis Graphs screen.

Other Graphs button goes to the Summary Graphs screen.

Print button is not functional. However, the print command on the Access menu can be used.

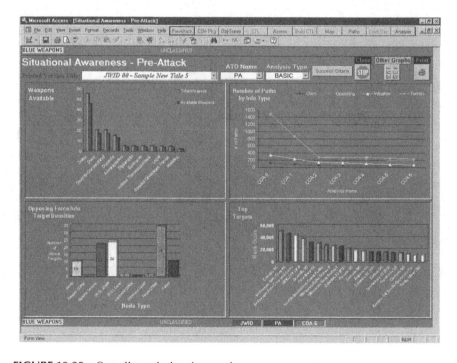

FIGURE 10.39 Overall attack situation graphs.

10.5.27 WEAPONS AVAILABLE

The Weapons Available screen on the Situation Awareness screen shows the total number of weapons and the total number of weapons available by category (Figure 10.40).

Click on the screen on the quad chart to enlarge it to full screen.

10.5.28 NUMBER OF PATHS BY INFO TYPE

The Number of Paths by Info Type screen on the Situation Awareness screen shows the total number of pathways for each information type (Figure 10.41).

Click on the screen on the quad chart to enlarge it to full screen.

Note: As analyses are conducted, this screen will update to show how each analysis has reduced the number of paths.

10.5.29 OPPOSING FORCE INFORMATION TARGET DENSITIES

The Opposing Force Info Target Densities screen on the Situation Awareness screen (Figure 10.42) shows the number of targets for each type of node used in the scenario. When enlarged (by clicking anywhere on the screen) it goes to the Info Target Densities – Pre-Attack screen that shows the four information requirement categories individually (Opposing Force, Own Force, Terrain, and Weather).

Click on the screen on any one of the quad charts to enlarge it to full screen.

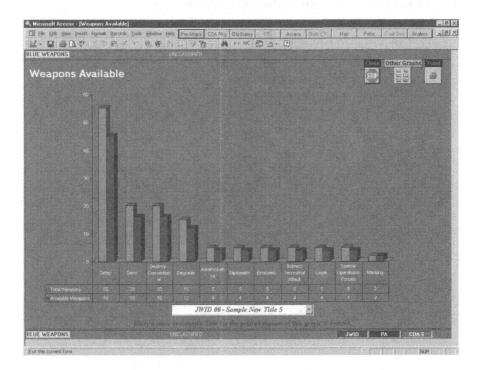

FIGURE 10.40 Weapons availability statistics.

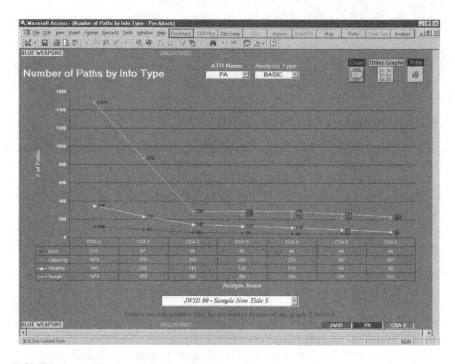

FIGURE 10.41   Total number of available network paths.

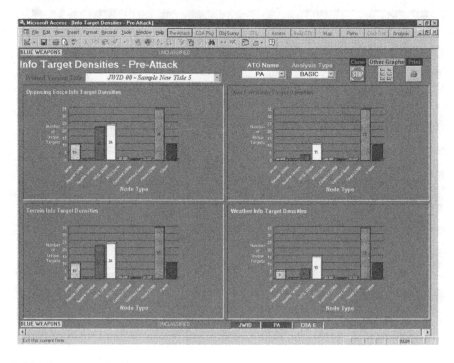

FIGURE 10.42   Information target densities.

Note: As analyses are conducted this screen will update to show reductions in the Target Densities. If there are no information pathways for a particular Information Requirement then no results will be shown.

10.5.30   Top Targets—Operations Menu

The Top Targets screen (Figure 10.43) on the Situation Awareness screen shows the top targets, ranked left to right by their relative score.

Click on the screen on the quad chart to enlarge it to full screen.

The Log button shows the same graph logarithmically.

The Target List button goes to a list view with more details on the top targets.

Note: As analyses are conducted, this screen will update to show changes in the relative scores of the targets.

10.5.31   Success Criteria (Success Criteria—Number of Paths)

The **Success Criteria** button on the Situation Awareness screen goes to the Success Criteria – Number of Paths screen (Figure 10.44). This shows a graph of the number of total paths and the success criteria level (shown by the red dot or red line) for each of the four information requirement categories (Opposing Force, Own Force, Terrain Info, Weather Info).

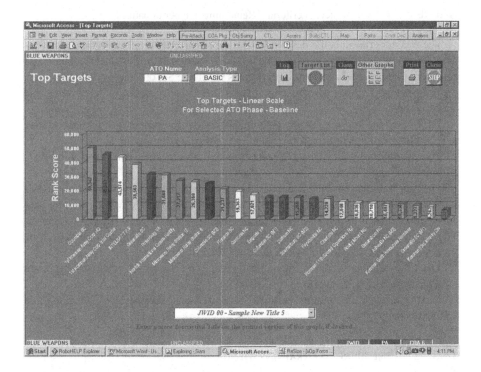

FIGURE 10.43   Top recommended targets to attack.

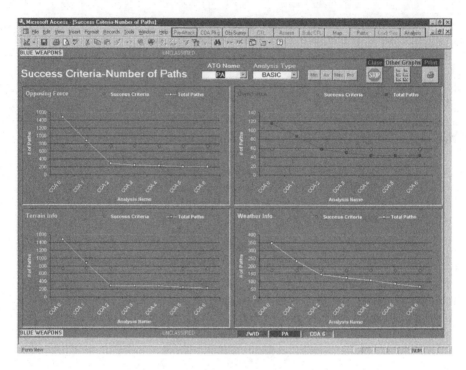

FIGURE 10.44   Attack success criteria—Remaining number of paths.

The success criteria is the desired level to be reached by eliminating pathways/targets. This is the level that has been determined to show that the desired effect has been achieved. The success criteria (red dot/line) is set on the Set or Delete Analysis screen.

The Success Criteria screens can also be sorted and viewed by ATO name and analysis type.

The Close/Stop button returns you to the Situational Awareness—Pre-Attack screen.

The Other Graphs button goes to the Summary Graphs screen.

The Print button takes you to a print preview screen. Then you can use the print command on the toolbar.

The four buttons shown in Figure 10.45 go to screens that show success criteria as it relates to time (showing if the information can be delayed along the pathway long enough to render it useless) for each of the four information requirement categories. The graphs show the time it takes for information to flow along the paths and the success criteria level (shown by the red dot or red line).

The time screens all have the same format. The Success Criteria—Maximum Path Delay screen is shown in Figure 10.45.

Each of these time screens can also be sorted and viewed by ATO name and analysis type.

10.5.32 Objectives—Operations Menu

These are summary screens. No inputs are made on these screens. They list the Objectives with their corresponding Priority, Success Criteria, and Paths

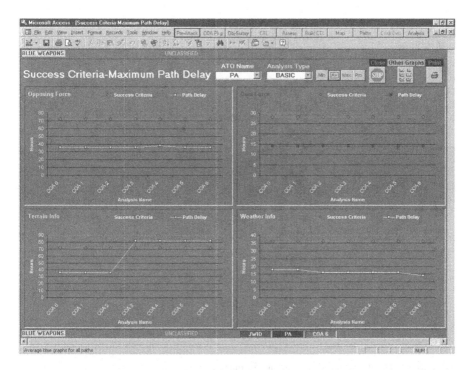

FIGURE 10.45   Attack success criteria—Induced time delays.

Available. These screens are cumulative and show the results of the last analysis that was completed (you cannot go back and see how a previous analysis affected the Objectives).

After an analysis has been run you can evaluate the effectiveness of the weapon applied against the node by checking the percent in the Paths Available column. All objectives start with 100 percent paths available, and then are degraded as pathways (nodes) are eliminated. A green background means the objective will probably still be accomplished. A red background means the objective will probably not be accomplished.

The best way to utilize this screen is to compare the results of a baseline run with no weapons applied against subsequent runs with weapons applied. This will allow you to observe the changes made by the elimination of specific pathways (nodes). See the Objectives Summary topic under Analyzing Results in the SIAM Operator Procedures chapter (Figure 10.46).

You can verify which results you are viewing by checking the objectives flow name at the top center of the screen.

Click on the Objectives button to go to the Objectives Full Summary screen (Figure 10.46). This screen shows the objectives from the national goals down to the command decision for both Blue and Red. The category, type, and number of each objective is shown.

The number of the record showing is at the top center of the screen just under the Objectives Flow Name field. Directly under that box is the Link Objective column. This is used to record the level at which the Blue and Red objectives intersect, or

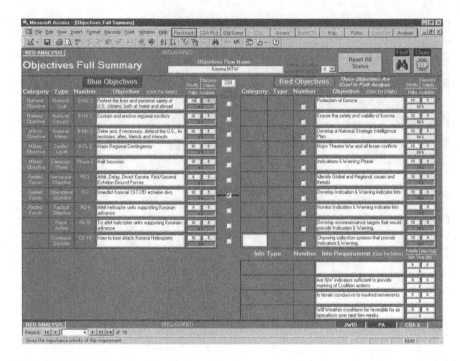

FIGURE 10.46   Objectives fulfillment summary.

directly affect, each other. Place a check in the box at the corresponding level and a black line will link the Blue and Red objectives at that level.

To view different records on the Objectives Full Summary screen:

Blue objectives: Use the Forward and Back buttons associated with the access record counter in the lower left corner of the screen. It will show you the current record that is displayed. This record number is also shown at the top center of the screen.

Red objectives: Use the Page Up/Page Down buttons on the keyboard. First you must click in any box/field on the Red side. This will activate the Page Up/Page Down buttons to work with the Red records. (The Page Up/Page Down buttons will also work to view the Blue records if you have clicked on that side. However, this is the only method that allows viewing of the Red records.)

The information requirements corresponding to the command decision are shown at the bottom of the screen. The inputs for the Priority, Min Prob, and Min Time (Hr.) fields are made on the Commander Decisions screen.

The Reset All States button will reset the Paths Available blocks to the default values as they would appear after a baseline analysis (for a Red analysis the Blue objectives would all be 0 percent and red, and the Red objectives would all be 100 percent and green). When this button is clicked, a pop-up dialogue box will appear to warn you and ask, "Are You Sure You Want to Reset All Objective States?" Only click "Yes" if you want to start over with default values in the Paths Available blocks.

To enter data on this screen (normally completed by the SIAM Analyst):

The inputs for these objectives (white fields) are made on the <u>Objective Level</u> screens. Click in any objective field (white field) to go to that screen (Objective Level).

10.5.33   PRIORITIZE CTL

This is the main operator screen (Figures 10.47 and 10.48). Here you can conduct analyses, view the prioritized targets, and select targets for the CTL. From this screen, you can also connect to other screens for many other functions that will be discussed.

Fields Containing Information on the Nodes/Targets:

The informational fields will be discussed in order from left to right as they appear on the Prioritize CTL screen.

Priority: Gives the order of the target list from highest to lowest priority. The first 25 targets are numbered. You can change the priority by selecting the appropriate number from the drop-down menu, or type in one.

Score: This is the target ranking score as calculated by SIAM in the previous analysis. It is a relative score that takes into account the importance of the information flowing through the node, the size of the choke point the node

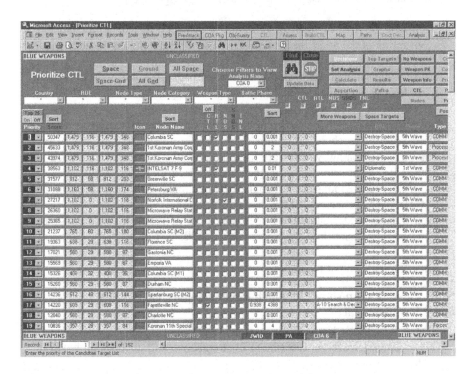

FIGURE 10.47   Candidate target list (CTL) assessment—left screen.

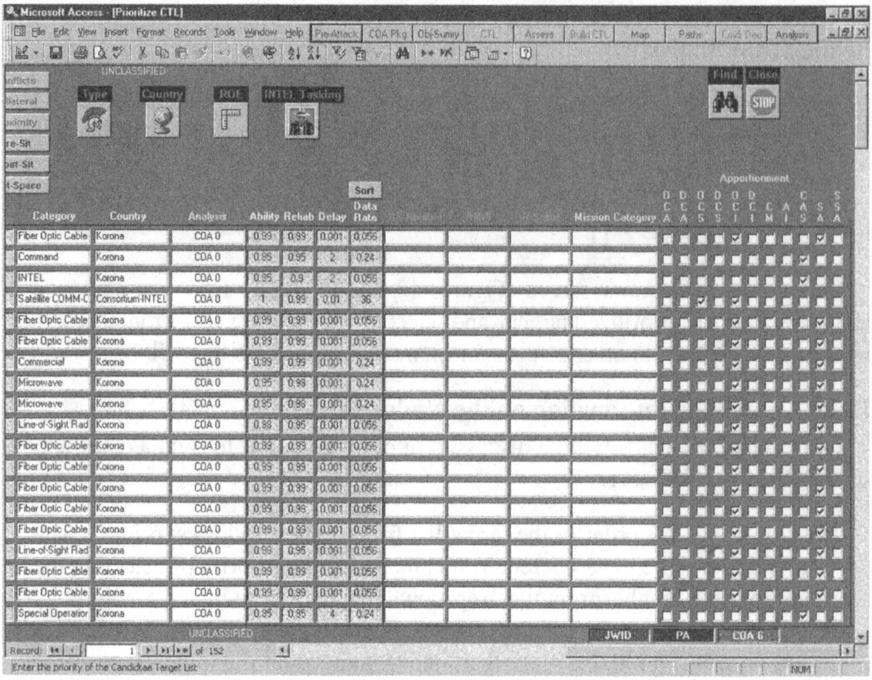

FIGURE 10.48   Candidate target list (CTL) assessment—right screen.

creates, the total number of nodes in that category, the capability and survivability of the node, the probability of use for the path that the node is on, and the overall attack strategy.

Choke point numbers: Shows the number of paths that the node is located on for the particular information category (Opposing Force, Own Force, Terrain, and Weather).

Note: There may be nodes listed after the prioritized targets that have no priority, score, or choke point numbers. These are nodes that are not in any information pathway. They are listed here so that they can be included in the CTL as targets. However, they are not used in analyses conducted in SIAM, and weapons are not applied against them. To include then on the CTL, place a check in the box in the CTL column.

Icon: Symbol to designate the three main types of systems (satellites, space related terrestrial sites, non-space-related terrestrial sites).

Node Name: Name of the node.

Node Classification shows in which target category each of the nodes has been placed:

BDA/ATO—Battle Damage Assessment/Air Tasking Order

CTL—Candidate Target List (the Off button above will uncheck all boxes in the column)

RTL—Restricted Target List (identified by the military category code from the MIDB; usually automatically checked if applicable)

NOS—No Strike List (identified by the military category code from the MIDB; usually automatically checked if applicable)

SNOS—SIAM No Strike List

TNL—Target Nomination List

Weapon Information—Gives information about the weapon system. This information is filled in when a weapon is selected from the drop-down list in the Weapon 1 column. It will show the combined values for 1 to 5 weapons employed against the target.

Total PK—Probability of kill of the target

Total Delay—Maximum time delay of information through the target

Total Pd—Probability of detection; does the target know that it has been attacked

Total Pa—Probability of attribution; does the target know the identity of the attacker

Weapon 1—Click on the drop-down arrow to see a list of weapons available and their capabilities. Click on the row the weapon is in to select it for use. You can apply more weapons by clicking on the More Weapons button above. This will expand the Weapon column to allow you to apply up to five weapons against a target.

ROE—Rules of engagement; this is the most severe type of weapon that can be applied against the target. It is based on the ROE for the country that owns the target. However, you are not prevented from applying a weapon with a greater severity than listed in the ROE.

10.5.34   Apportion (Set or Save ATO Phase)

Apportion button: Goes to the Mission Apportionment section of the Set or Save ATO Phase screen (see Figure 10.49). It shows which Air Force missions are supported by attacking a particular set of targets. Goals can be set for each mission type, and then as analyses are conducted these results can be compared to the goals.

The Goal section on the left lists all of the Air Force missions. For each mission type you can set a goal for the number of targets that you want to attack by entering that number in the field (use the arrows to scroll up or down). Those numbers will automatically be transferred to the bar graph to the right for the respective ATO goal. To change a particular ATO goal you must be on the record for that ATO.

The Results section below the goals reflects the number of targets by mission type that were targeted during a particular analysis (had a weapon applied against them). This is filled in automatically when an analysis is conducted.

The targets are automatically paired with the mission type(s) they fulfill according to the military category code that was listed on the MIDB for that node when it was imported into the SIAM database. These codes can be viewed on the node data entry screens, and the mission types that they fulfill can be viewed on the far right side of the Prioritize CTL screen.

Use the CTL button to return to the Prioritize CTL screen.

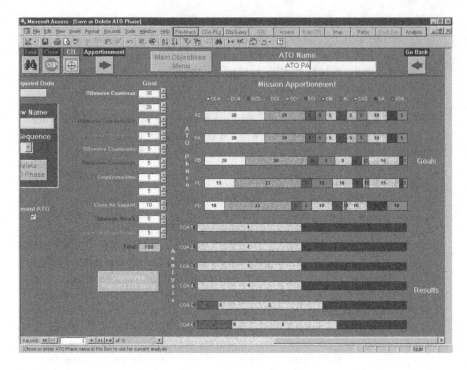

FIGURE 10.49   Mission apportionment by ATO phase.

10.5.35 POST-SPACE (SITUATIONAL AWARENESS—POST-ATTACK SPACE)

The **Post-Space** button goes to the Situational Awareness – Post Attack-Space screen (Figure 10.50). This set of four charts lets you view the current space situation as represented by the state of the weapons, information paths, target densities, and top targets, after analyses have been conducted.

This screen is basically the same as the <u>Situational Awareness—Post-Attack</u> screen. However there are two main differences:

1. The Space Weapons Employed screen shows only the space weapons used.
2. The Top Space Targets screen shows only space targets.

10.5.36 ASSESS CTL (SITUATIONAL AWARENESS—POST-ATTACK)

This screen lets you view the current situation, as represented by the state of the analyses that have been conducted. It will update and show the situation after each new analysis has been run.

This screen is basically the same as the <u>Situational Awareness – Pre-Attack</u> screen discussed earlier. Its screen is divided into the same four screens. The additional properties of these four screens will be discussed.

Click on any screen on the quad chart to enlarge it to full screen.

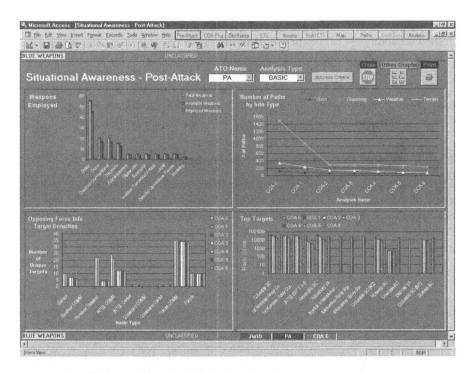

FIGURE 10.50 Post-attack situational awareness.

Weapons Employed—This screen also shows the employed weapons (used in an analysis) by category.

Number of Paths by Info Type—This screen updates to show an additional point for each analysis conducted, and this point is then connected by a line to the previous point. The buttons below show up on the enlarged screen.

Opp Pie button goes to the % Paths Denied—Opposing Info screen. This is a pie chart that shows the effect in percentage that each analysis had on the number of pathways that were denied.

Own Pie button goes to the % Paths Denied—Own Info screen. This is a pie chart that shows the effect in percentage that each analysis had on the number of pathways that were denied.

Ter Pie button goes to the % Paths Denied—Terrain Info screen. This is a pie chart that shows the effect in percentage that each analysis had on the number of pathways that were denied.

Wx Pie button goes to the % Paths Denied—Weather Info screen. This is a pie chart that shows the effect in percentage that each analysis had on the number of pathways that were denied.

All Paths button goes to the Total Number of Paths screen. This line graph shows the effect that each analysis had on the total number of pathways.

Opposing Force Info Target Densities—This screen updates after each analysis to show an additional bar on the graph for each node category, in a color assigned to that analysis. It shows a reduction in the number of targets, by type, for each analysis.

Top Targets—This screen updates after each analysis to show an additional bar on the graph for each target, in a color assigned to that analysis. It shows changes in the relative score for each of the top targets. If a target is eliminated by the analysis, then a bar will not be shown on the graph for that particular analysis.

10.5.37 Choke Points—Summary Graphs Screen

The Choke Points button goes to the Choke Points screen (Figure 10.51). It shows a graph of the nodes listed from left to right in order of the total number of times that node appears in the pathways of each of the four information requirements (Own, Opposing, Weather, and Terrain) depicted in separate colors. It also shows the total number of choke points for that node in all of the pathways combined in purple.

The 3D button goes to a screen that shows the same information as the screen in Figure 10.51, but it is a bar graph that is depicted in 3D.

The All button goes to the Probability of Kill by Weapon Type screen that shows the Pk of each of the weapon types in the database (Figures 10.52 and 10.53).

1. The following table reflects the data requirements of SIAM. A considerable effort was made to locate these data on INTELINK, but the hyperlink structure there was not conducive to downloading in database format.

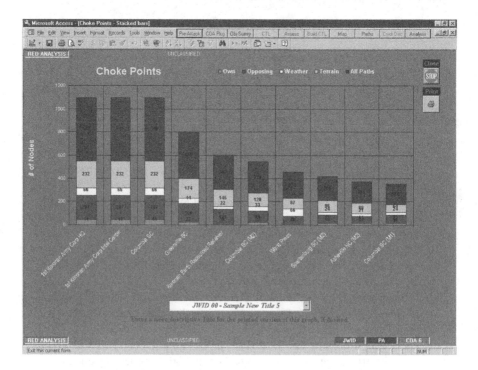

FIGURE 10.51 Network choke points.

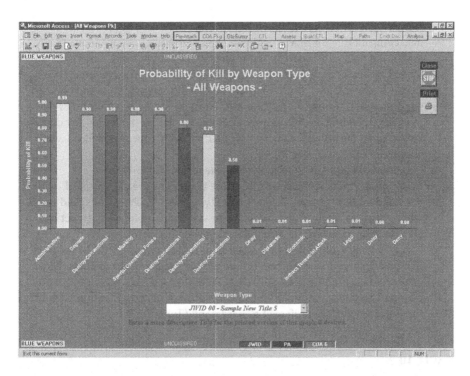

FIGURE 10.52 Probability of kill summary—all available weapons.

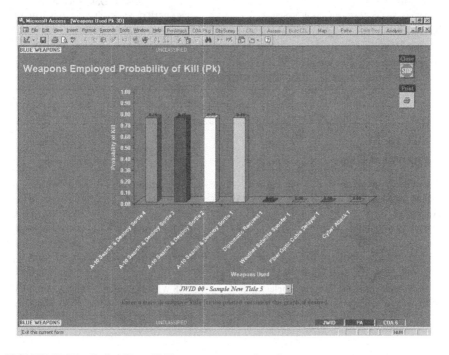

FIGURE 10.53 Probability of kill summary—employed weapons.

2. It is thought that this first data received will be a test of required fields and data compatibility for SIAM, and will be used to write code to automatically load these data into SIAM in the future. After compatibility verification, a formal database request can be made for permanent data synchronization. Because of continuing SIAM analyses both at the contractor site and government facilities, an update to these databases would be required every few months, depending on the analysis schedule. It is hoped that some way could be found to have these data on INTELINK for frequent download when required.

3. The data required falls into five main categories (Table 10.2):

4. In addition, Links & Nodes data for the country of Iraq is required, with the standard data output of node name, type, LAT, LONG, links, and data rates. Also, an unclassified list of node type code definitions (functional classification codes) in electronic format would be useful.

5. The following table reflects the document requirements of SIAM for data that probably can be obtained through local government channels. Most of these data are at the CONFIDENTIAL to SECRET level of classification. The latest versions of electronic documents/programs on CDs of these documents are preferred to hard copy (Table 10.3).

TABLE 10.2
Data Type Categories

Data Type	Possible Data Source
Satellites, Space Sensors, Launch Vehicles	SNAPSHOT, Mission Log, AIPSS, DIODE
Satellite Ground Receiver Stations, Launch Sites	SNAPSHOT, AIPSS, DIODE
Satellite Links	SNAPSHOT
Non-Space Ground Sites & Links	Links & Nodes, ADVERSARY
Non-Space Terrestrial Sensors (UAV, aircraft, SIGINT)	?

TABLE 10.3
Government Data

Document	Number	Source
JMEM Air-To-Surface Weaponeering System (JAWS) CD-ROM	61A1-3-11-CD	Mrs. Freda Rivers DSN: 336-5468
JMEM/AS AWOP Data Base	61JTCG/ME-3-19-2	Mrs. Freda Rivers DSN: 336-5468
JMEM/AS PC AWOP Data Base Foreign Weapons VS. U.S. Targets (Red-on-Blue)	61JTCG/ME-3-19-2	Mrs. Freda Rivers DSN: 336-5468
Military Intelligence and Security Supplement to the CIA World Fact Book (2 copies)	OSS WFS 97-001 or 1998 Version	Mrs. Ann Bartley (703) 482-0045

PART 3: EXAMPLE ANALYSES

10.6   HOW SIAM HAS BEEN USED IN THE PAST

The Space Warfare Center saw the potential of SIAM to perform future planning analysis and sponsored the tool to play in futures war games such as Blue-Flag 97-1; Joint Expeditionary Force Exercise (JEFX) 99, 02; JWID 00, 01, 02; Schriever I, II; and Navy Global 01. Also, SIAM has supported the following studies: QDR 01, Iraq (5 Analyses), Iran, North Korea, China, AFSCN vulnerabilities, Space Weather Network logistics optimization, and TENCAP SBR Info Distribution Timing.

The favorable warfighter's assessment of SIAM led to a HQ AC2ISRC/CC recommendation to CSAF that SIAM receive additional funding for full system development and subsequent fielding. This funding has led to the initiation of a complete rewrite of the tool in JAVA code using SQL server databases, using object-oriented software engineering and a spiral development process. This new version is called the Target Prioritization Tool for Links and Nodes (TPT-LN). A further refinement of the SIAM concept would link Blue, Red, and Gray forces information flows with military objectives and courses of action. This will allow the commander to see the effects of his possible actions and the reactions from the opposing side. All possible combinations of courses of action would be assessed for impacts on intelligence collection requirements.

Example analytic results for SIAM can be seen in Figure 10.54.

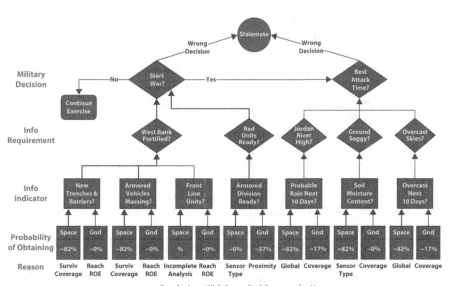

FIGURE 10.54   Example SIAM decision tree.

10.7 RESULTS

Examples of previous military wargames SIAM was employed in

1. *Navy Global 01 Wargame* (Newport, RI)—We participated in the Naval Global 2001 war game in Newport, RI from 12–27 July. We completed both Red and Blue SIAM database for the game country, and trained representatives of SWC/DOG to employ the model. After developing this database, we now have SIAM databases for all major threat countries available for all future government users. 80 SIAM analyses were accomplished during this war game showing the ISR and C2 vulnerabilities of the threat country. Ten major analytic briefings on these results were developed and presented to game participants. SIAM was well received as a very useful tool, and a new appreciation was given to the Navy for what space can bring to the conflict.
2. *JWID 01 Demo* (San Diego, Dahlgren, Norfolk, London)—Completed the development of the SIAM Blue and Red demo database for JWID 01. This will become the standard SIAM delivered test database. We participated in this demonstration from 9–27 July. The value of space to the warfighter was well illustrated by SIAM during all demo game runs and briefings.
3. *Space Weather* (Colorado Springs, Albuquerque)—We initiated a new SIAM study that will investigate space weather sensor and communications networks to determine reliability and maintainability sensitivities to outages. This is in support of Maj Quentin Dierks, Space Environment Support Systems, SMC Det 11/CIDA.
4. *Schriever 2001 Space Game* (Colorado Springs)—Participated in analyses supporting this war game using SIAM databases developed for AFSAA. Since this was the only links and nodes tool employed to the war game, SIAM was able to make a significant impact in showing the value of space to the war fighter. BGen Weston of the NRO directed us to provide a list of the top ten targets in the combined space and ground Red/Gray C4ISR networks. An integrated space-terrestrial targeting strategy was analyzed with SIAM, and statistical charts were generated to provide a convincing assessment of the most vulnerable centers of gravity on and off the battlefield. BGen Eliott, AIA/CV, LtGen MacGhey, AWC/CC, and MGen Looney, 14th AF/CC were also briefed these results. In addition, BGen Eliott requested SIAM support to develop a joint targeting scheme employing both conventional and CNA (IO) weapons.
5. *JWAC* (Dahlgren, VA)—Briefed and trained joint targeting staff on SIAM use. Developed a test scenario using their parameters to illustrate how SIAM may be used with their current databases. The hope is that SIAM will become part of their operational toolset.

10.8   OTHER SIAM ACTIVITIES

Other SIAM Non-Wargame Activities

1. *AFSAA* (Albuquerque)—Finalize QDR extension scenario databases when NAIC data have been received, and AFSAA has approved use of the Global Guardian data. We will then initiate analytic runs with this extended data, and submit these results to AFSAA.
2. *Joint C4ISR DSC* (Albuquerque)—The Joint Command, Control, Communications, Computers, Intelligence, Surveillance and Reconnaissance (C4ISR) Decision Support Center (DSC) is an organization within the Office of the Assistant Secretary of Defense for Command, Control, Communications and Intelligence OASD (C3I). They have requested that the SIAM model be registered into their C4ISR modeling and tools database. This has been accomplished, and the results will be online for DoD modeling and simulation community review.
3. *USSPACECOM IO Exercise* (Colorado Springs)—Participated in demonstrating SIAM at this exercise.
4. *Air Force Toolkit* (Albuquerque)—The Air Force Future Concepts & Development agency in D.C. has sent us their blue space databases. This is immediately being used to help develop future communications concepts for the Navy, to be used at Navy Global 01.

PART 4: EXAMPLE REAL-WORLD USE OF SIAM

Space weather (Colorado Springs, Albuquerque)—We completed the second draft of our SIAM analysis on the Solar Observing Optical Network (SOON) and Improved Solar Observing Optical Network (ISOON). The preliminary results were briefed to Col. Allen at AFWA on 11 December, and it was well received by him and his staff. They are currently looking for resources to support SIAM analytic efforts for their entire command. The final data for SOON/ISOON was entered into SIAM, and after multiple analytic runs, the results were briefed to SMC on 8 January 02.

We have completed our analysis of historical satellite coverage statistics for certain areas of the globe. We downloaded the latest NORAD two card orbital element sets, and employed PC SOAP as an orbital analysis tool. We then compared the orbits derived from PC SOAP with historical satellite coverage statistics for commercial satellite systems. We presented a briefing on the results, and re-wrote it at a lower classification level for electronic distribution.

We are completing our analysis of historical satellite coverage statistics for certain areas of the globe. We downloaded the latest NORAD two card orbital element sets, and employed PC SOAP as an orbital analysis tool. We then compared the orbits derived from PC SOAP with historical satellite coverage statistics for commercial satellite systems. We are now preparing a draft briefing on the results.

Target planning tool: Links and nodes (Colorado Springs, Albuquerque)—We continue to develop the SIAM follow-on: TPT-LN. The first draft of the software is completed. We conducted an interim demonstration for 14th AF on 16-17 October, and it was well received.

We twice researched documentation on this subject on INTELINK, and have obtained Space Surveillance Network documentation from MITRE and HQ AFSPC/DOYS. This information was then collated and standardized in an Excel spreadsheet, and 70 percent of it entered into PCSOAP for coverage and timing analyses. These data include military, civil, and worldwide commercial radars, optical telescopes, Tracking, Telemetry and Control (TT&C), and data readout stations.

Space imagery threat—We presented a briefing on the impact of commercial space imagery on military activities. This was based on work accomplished in previous reporting periods, and was presented to update new personnel on space threats.

- (U) Satellite coverage analysis—We completed an analysis of historical satellite coverage statistics for certain areas of the globe. We downloaded the latest NORAD two card orbital element sets, and employed PCSOAP as an orbital analysis tool. We then compared the orbits derived from PCSOAP with historical satellite coverage statistics data for commercial satellite systems, as provided by them in their marketing catalogs on the internet. We presented a briefing on the results, and re-wrote it at a lower classification level for electronic distribution.
- (U) Space situational awareness—We received training in Colorado Springs on the GFE tool, WebTas, that will enable timeline and process sequencing analysis of satellite launch, checkout, and status monitoring. A demo of this capability was being prepared. Also, we researched documentation for SSA on INTELINK, and have obtained Space Surveillance Network documentation from MITRE and HQ AFSPC/DOYS. This information was then collated and standardized in an Excel spreadsheet, and after obtaining the latest version of PCSOAP, these data were entered in for coverage and timing analyses. These data include military, civil, and worldwide commercial radars, optical telescopes, Tracking, Telemetry and Control (TT&C), and data readout stations. A demo of these analytic capabilities was presented, illustrating gaps in coverage for satellite status assessment.
- (U) Integration efforts—We conducted a preliminary analysis of what the interfaces may be for integrating diverse planning and control software. We developed concepts for uniting this coverage analysis, timing, intel data, and control software. This will allow the user to assess the overall situation, plan for a response, and monitor the results. We developed a data flow diagram illustrating detailed information exchanged by the various software packages. We estimated overall development costs, and presented all of these results.
- (U) AFOTEC brief—We presented SIAM capabilities to AFOTEC/TSE. This meeting was hosted by Sharon Nichols of the Test Support Directorate, C4ISR Branch. Some interest in SIAM was expressed. In addition, Capt. Tim Wolf of AFOTEC/Det 4 in Colorado Springs found out about this meeting, and called us expressing an interest in SIAM, and we briefed him in Colorado Springs in January. He became very interested in pursuing SIAM analytic efforts in Colorado Springs.

PART 5: FINAL SUMMARY

10.9   CONCLUSIONS

Main SIAM Characteristics

1. SIAM is an operational analysis tool useful for both operations planning and as a decision aid.
2. Tracks information flows from sensors to the end user (sensor to shooter).
3. Includes a nodal/network analysis capability.
4. Utilizes the Observe, Orient, Decide, Act (OODA loop) concept to determine relative worth of links and nodes in the information flow.
5. Aligns information requirements with military missions under a strategies-to-task structure.
6. Allows a direct correlation between information systems and military missions/tasks.
7. Allows the analyst to understand and quantify the impact of sensor, link, node degradation or loss on the military missions/tasks.
8. Allows the analyst to understand and quantify the impact of improving, modifying, or increasing the number of sensors, links, and nodes.
9. Could provide the link between logistics system metrics, Ao, MTBF, MRT, etc., and the overall mission the systems are supporting.
10. Answers the "so what" question—What happens (who cares) if I don't sustain, turn off, upgrade, etc., my sensors and sites?
11. This is something that any lead command should be doing to provide their commander a situational awareness for their mission and systems and to help make prioritization decisions

Section IV

Summary

11 Lessons Learned and the Proposed Way Ahead

Larry B. Rainey

11.1 CHAPTER 1

Chapter 1 is the introduction. There were no lessons learned and proposed way ahead statements.

11.2 CHAPTER 2: SYSTEMS TOOL KIT

11.2.1 LESSONS LEARNED

Systems Tool Kit (STK) was created in response to a group of engineers being tasked to do similar work on multiple projects. With each new project, they found themselves rewriting and tweaking code that they had already written for a previous project. STK was born when a few of these engineers realized how much time and effort could be saved by creating a mission engineering software tool that made common workflows simpler to implement and analyze. Throughout the decades, new features, workflows, and capabilities have been added to STK to make it a more robust tool. STK now provides insights to engineers and analysts at every level of a mission. It has been used to model systems throughout the design lifecycle, evaluate a system's effectiveness, diagnose and understand faults during operations, and simplify countless other tasks. As STK users provide feedback, features are added and capabilities are enhanced to make the tool more useful. In this way, STK is a dynamic tool that adapts to address the most pressing problems in aerospace and defense.

11.2.2 PROPOSED WAY AHEAD

Since STK is always evolving, it is challenging to say what will be added or enhanced next. With a current emphasis on model-based systems engineering (MBSE) in the industry, there is a desire to find the synergies between MBSE and STK and incorporate them with STK. Other new features and capabilities of interest are related to integrating STK with other analysis tools. For example, to better model a satellite's thermal state, a thermal analysis tool such as Ansys Thermal Desktop could be integrated with STK. STK could provide geometry information for where the Sun is with respect to a satellite and this information could be used in a rigorous thermal analysis tool to determine if the satellite is at risk of overheating a component. There are endless possibilities to integrate STK with other analysis tools; therefore, many efforts are being made to identify and connect these tools to provide users with enhanced capabilities.

DOI: 10.1201/9781003321811-15

11.3   CHAPTER 3: SYSTEMS EFFECTIVENESS ANALYSIS SIMULATION

11.3.1   Lessons Learned

System Effectiveness Analysis Simulation (SEAS) has proven remarkably successful and efficient, evolving into a robust simulation environment actively used today
by multiple services to assess the value of space to the fight. One of the key lessons
learned from 25 years of SEAS modeling, simulation, and analysis the importance
of multifaceted study teams able to delineate the primary mission linkages in how
the multi-domain force is fought.

11.3.2   Proposed Way Ahead

The next generation of SEAS, version 4.0, which is pending release at the time
of this writing, brings major upgrades and modernization to both the engine and
user interface. The ability to execute Python code and its formidable libraries from
inside the SEAS Tactical Programming Language (TPL) are opening new horizons.
Future advancements in SEAS will rapidly move toward closing the Model-Based
Systems Engineering (MBSE) loop via integration of SysML block definition and
state machine diagrams as part of routine modeling, simulation, and analysis support
to space community.

11.4   CHAPTER 4: SATELLITE ORBIT ANALYSIS PROGRAM

11.4.1   Lessons Learned

11.4.1.1   The Long and Winding Road

Since the porting of the SOAP precursors to the MS-DOS-based personal computer (PC) platform in 1988, the software has been under development for 35 years.
During this time, the focus has been on improving capabilities, performance, and
portability. Programming for performance using memory and graphics hardware of
the early days was a challenge. As computing has evolved from 8 to 16 to 32 to 64
bits, more breathing room has been provided for less convoluted, but increasingly
more resource-intensive approaches. However, the residual effects of the early bit
and byte counting can still be seen in the SOAP architecture to this day.

Initially, custom interfaces were developed for the native windowing environments for the PC, Macintosh, and Sun Workstations. This was carried forth though
the transition from DOS to Windows on the PC, from Motorola 68000 to PowerPC to
Intel to M1 on the MacIntosh, and from Unix to Linux on workstations, with a pas-
sage through many different desktop environments. The native graphics APIs were
unified into a single OpenGL interface across all platforms in the late 1990s. The Qt
Application Framework was adopted for a portable user interface in the early 2000s,
after a brief stint with a multi-platform framework called Zinc.

Once the portable framework was in place, the burden in supporting three disparate computing platforms has lessened considerably, but each platform continues to
have its strengths, weaknesses, and issues.

11.4.1.2 The Persistence of Obsolescence and the Least Common Denominator

The rapid growth in computer technologies over the last 30 years presented unique challenges to software developers, SOAP team included. On the Macintosh platform, SOAP was caught in the growing pains of a computer company in transition. "Carbon compatible," "Big Endian," "Fat Binary," and "Universal Binary" became familiar terms as the company switched APIs, microprocessor families, and macOS kernels.

On the Windows platform, perhaps the greatest lesson learned in the development and deployment of SOAP is that end users and their administrators do not keep their systems up to date, or even in good repair. This is especially true of OpenGL drivers, upon which SOAP depends. There are often cases of obsolete, missing, or incorrect configurations, particularly on systems not connected to the internet. In some cases, this is intentional, as systems are under configuration control and difficult to reconfigure and update. Overall, graphics intensive programs such as SOAP are a very small minority of applications, and one that administrators are ill-equipped to support. The situation has been improving with modern systems, but with the onset of virtual environments, we are reliving these issues all over again.

Beginning with the pandemic, users began accessing their workplaces remotely though environments such as the Citrix Workspace. In 2020, these did not have adequate OpenGL support for running SOAP. As a result, the developers at The Aerospace Corporation identified the OpenGL MESA libraries as a capability that could provide software emulation of OpenGL in virtual environments. These Open-Source libraries were thus deployed with SOAP and their availability documented. We did not want these to be the default configuration, as software emulation degrades performance on systems with good hardware drivers. Instead, we provide the MESA libraries in supplemental directories with instructions on how to mitigate failures in readme files. It has taken administrators in many government and contractor administrations three to four years to recognize such shortcomings in their virtual environments and deploy the needed updates. We still must field related technical support inquiries to this day.

Given this convoluted landscape, it is important to have a resilient application that fails gracefully and without crashing. For example, if graphics shaders are not supported, SOAP will run without them at reduced visual fidelity. If the graphics drivers are completely absent, the SOAP won't get past the splash screen, but will refrain from crashing and its user interface will remain functional. There is an option in SOAP to present the driver information to the user, so this can be reported to the developers.

11.4.1.3 Portability or the Lack thereof

Portability is less of a problem today than it used to be, but there is still some reliance on architecture-specific aspects of the MS-Windows, macOS, and Linux platforms. With the widespread use of Intel processors, nearly all systems now use Little Endian byte ordering, although the code in SOAP for detecting and converting Big Endian data is still present.

There are still differences across file systems, such as the use of the back slash, '\', versus the forward slash '/' and the naming of volumes. SOAP scenarios have many ancillary files that must be referred to in the master scenario file. Sometimes scenarios are exchanged across architectures and can be problematic even on the same architecture. For example, on MS-Windows, I may have an "E:" drive, but a colleague may not. SOAP handles this by storing the referenced file names only and omitting the paths and volumes. There is a specific search order through which external files are located and loaded:

1. The scenario directory
2. The designated supplemental directory
3. The "well-known" directories
4. The SOAP application directory

The first option is the most robust and commonly used, but it is not appropriate for all situations. There are cases where a common set of potentially large files are used by multiple scenarios or applications. In the case of a common collection of terrain or imagery, a Supplemental Directory can be used and is specified in the SOAP preferences, directory option. Separate paths are specified for each supported file type on Windows, macOS and Linux, so that saving the scenario on one architecture still preserves the settings for others. The "well-known" folders are a collection of locations for each file type installed relative to the SOAP application. Folders are included for items such as world maps, star maps, ephemeris files, images, and CAD models. When SOAP is deployed on a server, the user may not have write-access to these, and thus the other methods can supplant this. Finally, the SOAP application directory is used when the SOAP application is set up as part of a "pack and go" for another user, or the SOAP viewer.

11.4.1.4 Language Specifications and Software Libraries

Language and library issues are most acute on the Linux platform, where SOAP relies on resources in the native runtime environment. Unfortunately, a version of SOAP compiled for one distribution and version of Linux does not often run (and sometimes, doesn't even compile) on another. There are numerous distributions of Linux deployed throughout the community, such as Red Hat, Ubuntu, and Scientific Linux. Each distribution has different versions such as 6, 7, and 8 for CentOS, and 18, 20, and 22 for Ubuntu. Each combination of distribution and version potentially has different versions of the C++ standard implemented. The problem is not so much that SOAP doesn't compile, but rather some of the third-party libraries such as the Qt Application Framework do not. For example, on CentOS 7, we are locked into Qt Version 5.12, because the Qt version 5.15.2 used in our main build will not compile because it employs newer features of the C++ standard. Thus, we must conditionally compile some portions of the SOAP code that would normally rely on such features. We also must maintain a collection of virtual machines, one for each supported distribution and version of Linux.

Although modernists may disagree, the solution for SOAP has been to avoid cutting edge language constructs, and emerging features of APIs and drivers. The sweet spot for large scale deployment across a heterogeneous set of systems and user environments seems to be the use of technology that is about five years old. Although newer graphics APIs such as Vulkan offer enticing features and performance, support for this is not ubiquitous across the space systems community. The SOAP developers also studiously avoid proprietary and/or non-portable APIs such as Direct-X and gaming engines.

11.4.1.5   Standards or the Lack thereof

Different organizations employ different terminology and conventions for referring to astrodynamics concepts and geospatial data. Reconciling this requires that this be recognized and mitigated.

11.4.1.5.1   Definition of Earth-Centered Inertial (ECI) Coordinates

One foundational issue involves the definition of the vernal equinox, and the Earth Centered Inertial (ECI) coordinate frame. This frame is widely used for defining the location of Earth-orbiting objects. Originally identified as a relatively invariant frame, it undergoes differences in orientation in accordance with what convention is used. Options for defining ECI include

- J2000 (fixed)
- Mean equator mean equinox (precession)
- True equator true equinox (nutation)
- True equator, mean equinox (used by NORAD)

A full analysis of these is beyond the scope of this discussion, but in J2000 the motion of the equator and equinox is frozen at a reference time on January 1, 2000. Precession is a phenomenon where the direction of the vernal equinox is drifting slowly with respect to stellar references, nutation is the periodic wobble of the Earth's equatorial plane caused by the gravity of the Moon. The problem with such conventions is that different choices result in different positional values that can affect analysis results. Moreover, the use such conventions are implied within an organization, but are often not communicated between organizations. Making things more confusing, the frames in motion can be referenced either at a fixed reference time (at epoch) or continuously (of date). When publishing or exchanging coordinates, it is important to share the conventions being used.

11.4.1.5.2   Geospatial Considerations

One can be surprised at how many ways there are to define ground coordinates using latitude and longitude. There is the question of which one is presented first, and whether to use decimal degrees or degrees, minutes, and seconds (DMS). In SOAP, decimal degrees is preferred. Even though use of DMS is still pervasive, there are simply too many permutations of the format for it to be reliable. Using minutes and seconds as units of position can be confusing when juxtaposed with descriptions of time.

Also confusing is the use of the term "elevation" for both angles and heights. When referencing topocentric coordinates, the terms "azimuth," "elevation," and "range" are used, but if distance along the zenith vector is considered positive, then these form a left-handed coordinate system. In SOAP, coordinate systems are right-handed, and the terms "clock" and "cone" are used for positioning angles. The cone angle is measured off a defined pointing direction, and the clock angle is projection of a reference direction in the plane normal to the pointing direction. The advantage of clock and cone angles are that their use is intuitive in the context of an instrument and independent of any geospatial coordinate frame.

11.4.2 Proposed Way Ahead

It should be noted that the SOAP software is not developed proactively as a deliverable. The program was originally developed by The Aerospace Corporation for internal use. However, as government organizations became aware of SOAP, they engaged in a concerted effort to obtain the software. Because the government has unlimited rights to the software developed under contract, the corporation does provide it to their users and authorized contractors. Requests for SOAP are vetted and approved by Space Systems Center at Los Angeles Air Force Base. SOAP is export-controlled under the Export Arm Regulations (EAR99).

The advantage of developing SOAP in the non-profit Federally Funded Research and Development Center (FFRDC) environment is that there are many subject matter experts (SMEs) already on site to confer with. The disadvantage is that because the software is not a deliverable, there is no long-term road map, and development is performed in reactive spontaneity in accordance with the problems of the day. The current trends exerting influence on software development are support for analyzing the emergence of the large low Earth orbit constellations, such as Starlink, and the advent of cloud computing.

Modeling systems with large numbers of satellites requires that software convey an economy of effort in dealing with many objects. This affects data acquisition, workflow, performance, and the presentation of results. To some extent, this is an industry-wide problem in that the developers of the space catalog did not envision this multitude of objects with its 5-digit identifiers. The recalibration of the catalog has implications that will reverberate throughout the space community. SOAP has been programmed to support the emerging alpha-5 nomenclature while retaining compatibility with the legacy IDs. It has also been calibrated to support the bulk definition and analysis of entire groups of objects.

The impact of digital engineering (DE) and cloud computing is still playing out. Security considerations make distributed computing somewhat difficult. There are potential network bandwidth issues associated with the transmission of high-performance graphics. Finally, there is the heterogeneity of end-user systems and environments. These considerations conspire to create the likelihood that SOAP will not become a Cloud-based application in the near term. However, the potential of the cloud can be realized as a data source. Ideally, space objects, geospatial data, and real-time weather will all become accessible through this medium.

To conclude, we must come to grips with the graying of the aerospace industry. To keep moving forward, we need a new generation of talented developers to take the helm. In an era of declining benefits, it is difficult for us to earn the longevity and loyalty of the newer staff that we ourselves were able to both obtain and provide. We are relying on the wisdom of our leadership to make it so.

11.5   CHAPTER 5: ARCHITECTURE FRAMEWORK FOR INTEGRATION, SIMULATION, AND MODELING

11.5.1   Lessons Learned

1. AFSIM is both an engagement and mission-level simulation.
2. AFSIM was developed to address analysis capability existing in legacy simulation environments.
3. AFSIM can simulate missions from subsurface to space and across multiple levels of simulation model fidelity.

11.5.2   Proposed Way Ahead

AFSIM should have a capability for a deeper dive into how various space capabilities can support both engagement and mission-level simulations that in turn reflect real world operations for air, ground, and sea operations.

11.6   CHAPTER 6: EXTENDED AIR DEFENSE SIMULATION (EADSIM)

11.6.1   Lessons Learned

1. EADSIM is a system-level simulation of air, space, and missile warfare developed by the U.S. Army Space and Missile Defense Command Space and Missile Defense Center of Excellence's Capability Development Integration Directorate.
2. EADSIM provides an integrated tool to support joint and combined force operations and analyses.
3. EADSIM is also used to augment exercises at all echelons with realistic air, space, missile, and Battle Management, Command, Control, Communications, and Intelligence (BM/C3I) warfare.
4. EADSIM is used by operational commanders, trainers, combat developers, and analysts to model the performance and predict the effectiveness of ballistic missiles, surface-to-air missiles, aircraft, and cruise missiles in a variety of user-developed scenarios.
5. EADSIM is one of the most widely used simulations in the Department of Defense.

11.6.2   Proposed Way Ahead

EADSIM is a system-level simulation of space and other venues that supports joint and combined force operations and associated analyses. EADSIM is also used to augment exercises at all echelons with realistic space and other resources for warfare analysis. More detail could be provided as to how space is simulated to address a customer interested in how space supports the end customer.

11.7 CHAPTER 7: OVERVIEW OF SYNTHETIC THEATER OPERATIONS RESEARCH MODEL

11.7.1 LESSONS LEARNED

1. STORM is the primary campaign analysis tool used by the Office of the Chief of Naval Operations, Assessment Division OPNAV N81 and other Department of Defense organizations to aid in providing analysis to top-level officials on force structures, operational concepts, and military capabilities.
2. This overview of STORM analyzes the variability associated with many replications and evaluates the trade-offs between the expected number of replications and the precision and probability of coverage of confidence intervals.
3. The results of this research provides OPNAV 81 with the ability to capitalize on STORM's full potential on a timeline conducive to its high-paced environment.
4. The distribution of outcomes is examined via standard statistical techniques for multiple metrics. All metrics appear to have sufficient variability, which is critical in modeling the combat environment.
5. The trade-off for confidence intervals between the expected number of replications, precision, and the probability of coverage is very important.
6. If a more precise solution and a higher probability of coverage are required, more replications are generally needed. This relationship is explored and a framework is provided to conduct this analysis on simulation output.

11.7.2 PROPOSED WAY AHEAD

It is stated that STORM is a closed-form analytic simulation of air, space, ground, and maritime simulation. In the discussion, naval assets and air assets are addressed but not space assets. STORM should also address how space is modeled and how it supports a decisionmaker.

11.8 CHAPTER 8: SPACE WARGAMING: AN INTRODUCTION AND CYBERSECURITY AS IT RELATES TO WARGAMING

11.8.1 LESSONS LEARNED

The Space Training and Readiness Command was first established on November 1, 1993, as the U.S. Air Force Space Warfare Center (SWC) under Air Force Space Command and simply renamed to the Space Warfare Center (SWC) on July 1,1994, as the unit was stood up at Schriever Air Force Base, Colorado. On March 1, 2006, the SWC was redesignated as the Space Innovation and Development Center (SIDC).

The SIDC was inactivated on April 1, 2013, and its responsibilities were transferred to Air Combat Command's United States Air Force Warfare Center. The establishment of the Space Warfare Analysis Center (SWAC) was ordered by Chief of Space Operations John W. Raymond. Originally planned as Space Warfighting Integration Center, Vice Chief of Space Operations David D. Thompson was tasked to focus on its establishment upon taking office. Raymond approved the organizational design of SWAC on March 8, 2021. It was activated on April 5, 2021, in Colorado Springs, Colorado.

11.8.2 Proposed Way Ahead

The SWAC needs to consider the effects of emergence both positive and negative upon space system of systems. Positive can be a force multiplier and negative can be a significant detriment. This needs to be especially looked at in the context of threat considerations. The results have a significant impact upon force design considerations. SWAT might be a good laboratory to examine these effects. The bottom line may be: how secure or not are our satellites to a cyber-attack on orbit (i.e., the threat).

11.9 CHAPTER 9: SPACE WARFARE ANALYSIS TOOLS (SWAT) AND SPACE ATTACK WARNING (SAW)

11.9.1 Lessons Learned

Space object change detection in SWAT appears to be able to automatically determine when a space object, whether live or dead, has changed in some way (orbital or characteristics), and present to the user a prioritized list for further space surveillance/SOI assessment (for both Red and Blue space objects). It also appears that space object optical data can have a significant impact on this state change algorithm. The optical data appear to allow distinguishing satellite types and significant characteristics differences. The SWAT choke point/SAW maps have generally been well-received by the space community as a unique way to assess the space situation. They deserve further development. The intelligence indicators data developed for the SWAT Red Courses of Action (COA) detection algorithms have been well received by the community and extensively used by AFRL/RIE contractors and proved critical for program success. The actual ability to detect Red COA can only be fully proven in an actual space war. However, space war simulations, wargames, and exercises can help determine its validity in the future.

11.9.2 Proposed Way Ahead

11.9.2.1 Operationalize Space Object Change Detection Algorithms

SWAT Space Object Change Detection appears to be working and providing significant data on a bi-weekly basis. This data should be passed to the combined space operations center operators to help them in prioritizing their space surveillance/SOI requirements.

11.9.2.2   Obtain Space Object Optical Sensor Data

Space object optical sensor data had a significant impact on the space object change detection algorithms in SWAT. However, this data was mostly on geosynchronous satellites, and the latest data measurement date was in 2007. More recent and extensive data are required. For example, Los Alamos National Labs has ground-based optical data, and somewhere there exists previous SBV space-based sensor that may provide MEO and LEO data. The STSS program can provide additional, current data. Finally, the GEODSS system may have optical data measurements that can be obtained.

11.9.2.3   Evolve Space Choke Point Maps

The SWAT choke point/SAW maps have already proven useful during ACE exercise testing. However, this was just the beginning, and additional displays need to be designed. For example, the AVIS display shows when space objects have similar altitudes and inclinations, making them "close" in orbital space in terms of the amount of delta-v required to rendezvous with the potential target. In the case of the COSMOS 2251 and Iridium 33 collision[1] which had a 12-degree difference in inclination, but the same orbital altitude, this would be displayed on a common horizontal line in AVIS. Obviously, a better way must be developed to illustrate "glancing" attacks against satellites where the inclination does not have to be the same for both objects. Also, it is intended to design delta-v versus transit time maps in some way on these charts.

11.9.2.4   Develop Space Wargaming Toolsets

The best way to test the SWAT choke point/SAW maps and the Red COA detection algorithms is in theoretical space wargames. SWAT already has automatic space scenario generation tools, so these need to be expanded to include a full wargame testing capability. This would include tools to help develop strategies, and optimize delta-v versus time during synchronized, multiple attacks employing weapon systems of varied phenomenologies. This would test possible space attack envelopes, sequences, and tempos while training future warfighters in detecting surprise attacks in space. Ultimately, this provides Red and Blue optimal attack strategies based on SID (SatAC Satellite Information Database) satellite characteristics and sensor orbital data.

11.10   CHAPTER 10: SPACE INFORMATION ANALYSIS MODEL TOOL

11.10.1   Lessons Learned

SIAM is an operational analysis tool useful for both operations planning and as a decision aid because it tracks information flow from sensors to the end user (sensor to shooter). It also utilizes the Observe, Orient, Decide, Act (OODA loop) concept to determine relative worth of links and nodes in the information flow. This would be critical for commanders to evaluate whether to target the information space systems generate or deny the flow of this information on the ground. Thus, it provides a joint space-terrestrial target evaluation tool employing common measures of merit. It allows a direct correlation between information systems and military missions/tasks.

It can also provide the link between logistics system metrics, Ao, MTBF, MRT, etc., and the overall mission the systems are supporting. It answers the "so what" question—what happens (who cares) if I don't sustain, or turn off, upgrade, etc., my sensors and sites?

This is something that any lead command should be doing to provide their commander a situational awareness for their mission and systems and to help make informed prioritization decisions.

11.10.2 PROPOSED WAY AHEAD

This SIAM tool should be modernized and ported into major command and control systems to enable timely assessments of battlefield actions and results. It is currently in the desktop version of Microsoft Access and should at least be ported over into Microsoft SQL Server for analysis of larger databases. Since this tool can be beneficial to both terrestrial and space conflicts, it should be made available to command decision centers for both fields of operations. In addition, when planning for joint warfare operations, this critical tool can provide joint targeting of information flowing on the battlefield, whether though space systems or terrestrial communications and ISR systems. SIAM also provides common targeting technologies not only for United States military operations, but also for allied ones for combined operations that also include information-warfare planning. Finally, SIAM relies on military databases of links and nodes of sensor-to-shooter networks, so SIAM should be designed to automatically import these updated databases before calculating the best information paths to attack on the battlefield.

NOTE

1 Space.com, "U.S. Satellite Destroyed in Space Collision," Becky Iannotta and Tariq Malik, web site: http://www.space.com/news/090211-satellite-collision.html [11 February 2009].

Index

Note: Page numbers in **bold** and *italics* refer to tables and figures, respectively.

Printed in the United States
by Baker & Taylor Publisher Services